Array and Wearable Antennas

The text highlights the designing of efficient, wearable, and textile antennas for medical and wireless applications. It further discusses antenna design for the Internet of Things, biomedical, and 5G applications. The book presents machine learning and deep learning techniques for antenna design and analysis. It also covers radio frequency, micro-electromechanical systems, and nanoelectromechanical systems devices for smart antenna design.

This book:

- Explores wearable reconfigurable antennas for wireless communication and provides the latest technique in term of its structure, defective ground plane, and fractal design.
- Focuses on current and future technologies related to antenna design, and channel characterization for different communication links and applications.
- Discusses machine learning techniques for antenna design and analysis.
- Demonstrates how nano patch antennas resonate at multiple frequencies by varying the chemical potential.
- Covers the latest antenna technology for microwave sensors, and for fiber optical sensor communications.

It is primarily for senior undergraduate, graduate students, and academic researchers in the fields of electrical engineering, electronics, and communications engineering.

Array and Wearable Antennas
Design, Optimization, and Applications

Edited by
Puran Gour
Nagendra Singh
Rajesh Kumar Nema
Ravi Shankar Mishra
Ashish Kumar Srivastava

CRC Press
Taylor & Francis Group
Boca Raton London New York

CRC Press is an imprint of the
Taylor & Francis Group, an **informa** business

Designed cover image: Shutterstock

First edition published 2024
by CRC Press
2385 NW Executive Center Drive, Suite 320, Boca Raton FL 33431

and by CRC Press
4 Park Square, Milton Park, Abingdon, Oxon, OX14 4RN

CRC Press is an imprint of Taylor & Francis Group, LLC

© 2024 selection and editorial matter, Puran Gour, Nagendra Singh, Rajesh Kumar Nema, Ravi Shankar Mishra and Ashish Kumar Srivastava; individual chapters, the contributors

ISBN: 978-1-032-59234-3 (hbk)
ISBN: 978-1-032-72761-5 (pbk)
ISBN: 978-1-003-42244-0 (ebk)

DOI: 10.1201/9781003422440

Typeset in Sabon
by SPi Technologies India Pvt Ltd (Straive)

Contents

Editor's biography

Professor Puran Gour specializes in the research and development of patch antenna design and radio systems and works as a professor in the NRI Institute of Information Science and Technology Bhopal (M.P) in the Department of Electronics and Communication Department. For several years, he undertook research in microstrip patch antenna design and has more than twenty years of experience in the field of engineering institute. His current research is in antenna design and he focuses on sensor surveillance using both active and passive sensor devices. He has worked in various positions such as head of department, academic dean, and exam superintendent. He has published many research articles in reputed refereed journals and Scopus index IEEE conferences. He has also published three book chapters and authored one book for international publishers. The holder of four design patents and has also organized many national and international conferences as well as faculty development programs. He has also supervised more than twenty dissertations at the PG level. He is a member of Institute of Electrical and Electronics Engineers (IEEE) and the Institution of Electronics and Telecommunication Engineers (IETE) and also holds a lifetime membership of te Indian Society for Technical Education (ISTE).

Dr. Nagendra Singh is working as a principal of the institution and associate professor in the Department of Electrical Engineering at Trinity College of Engineering & Technology, Karimnagar, Telangana, India. He has about twenty years of experience in the field of teaching in engineering institutions. He has worked in various positions such as the head of department, academic dean, and exam superintendent. He has published many research articles in reputed refereed SCI and Scopus index Journal and IEEE conferences. He has also published 13 book chapters and edited 5 books published by Indian and international publishers. He has also been granted 14 design patents and has organized many national and international conferences as well as faculty development programs and seminars. His areas of interest include optimization techniques, Artificial Intelligence, power systems, renewable energy systems, power electronics, smart antenna, and blockchain technology. He is a life member of various professional bodies and also a reviewer of many international journals.

Dr. Ravi Shankar Mishra is professor in the Sagar Institute of Science & Technology (SISTec) Bhopal. He received his PhD in 2011 and MTech. in 2005 from N.I.T. Bhopal. He has completed his PG Diploma in VLSI Design from C-DAC Bangalore. He has almost two decades of research and academic experience in reputed institutions such as Lovely Professional University (LPU) Punjab, Symbiosis University of Applied Sciences (SUAS), Indore, and Guru Nanak Institute of Technology (GNIT) Hyderabad. He has handled important positions of responsibilities such asdean, head of department, research coordinator, etc. He has published more than 40 research papers in reputed international journals and international conference proceedings, including SCI, and Scopus. He is also a reviewer for many journals. One copyright and two books are registered with his name. He has supervised 3 PhD scholars and 18 dissertations at the PG level. He is a Life Fellow Member of the Institution of Electronics & Telecommunication Engineers (IETE) and the International Association of Engineers (IAENG). His research contributions have been in the area of designing efficient PUF-based circuits to using Complementary Metal-Oxide-Semiconductor (CMOS) for generating secure keys.

Professor Rajesh Kumar Nema specializes in research and development intomicrostrip patch antenna design and Radio Frequency (RF) devices and working as a professor in a IES College of Technology, Bhopal (M.P) in the departments of Electronics and Communication. For a number of years, he undertook research in microstrip patch antenna design and he has more than 21years of experience in the field of engineering institutes. In his career, he has achieved silver certification in an Antenna Designing course conducted by NPTEL IIT Bombay. His current researches are in antenna design and focus on the design and development of RF devices and antennas for 5G and 6G technology. He has published many research articles in reputed refereed journals, SCI and Scopus index IEEE conferences andhe has also published one book with an international publisher. He has also published three patents. He has organized many national and international conferences as well as faculty development programs.

Dr. Ashish Kumar Srivastava is currently working as a professor in Galgotias University Greater Noida in the department of Computer Science Engineering (CSE). He has more than two decades of teaching and research experience. He has completed his PhD in IT from MANIT Bhopal, his MTech.(IT) fromTezpur Central University Assam and his BE(EEE) from Bharathiar University Coimbatore Tamil Nadu. His area of research is computer networks, the Internet of Things (IoT) and its application, cybersecurity, wireless sensor networks (WSN), software-defined networking (SDN), and natural language processing (NLP). In his career to date, he has given valuable guidance to more than 23 MTech. students and more than10 BTech. students. He has filed nine patents and published 20 conferences and 26 journal papers, of which 5 are in SCI journals and some are in reputed journals and IEEE conferences. He is also a reviewer for many reputed IEEE conference and journals and is an active member of ACM, IEEE, and CSI society.

Contributors

Salini Abraham
Department of ECE
Jaibharath College of Management
    & Engineering Technology
Perumbavoor, Kerala, India

Jagdeesh Kumar Ahirwar
Department of Electronics and
    Communication Engineering
Gyan Ganga Institute of Technology
    and Sciences Jabalpur
MP, India

Priyamvada Chandel
Engineering Officer 4
Energy Meter Testing Laboratory
    Central Power Research Institute
Bhopal, MP, India

G Dhanya
Department of ECE
MGM College of Engineering and
    Technology
Pampakuda, Kerla, India

Madhukar Dubey
Department of Information
    Technology
ITM College
Gwalior, India

Rahul Dubey
Department of Electronics &
    Communication Engineering
MITS
Gwalior, India

Puran Gour
Department of Electronics &
    Communication
NIIST
Bhopal, India

Hemant Kumar Gupta
Department of Computer Science &
    Engineering
SAGE University
Indore, India

Megha Gupta
Department of Electronics &
    Communication
SAGE University
Indore, India

Pritesh Kumar Jain
Department of CSE
Shri Vaishnav Vidyapeeth
    Vishwavidyalaya
Indore, India

Sandeep Kumar Jain
Department of CSE
Shri Vaishnav Vidyapeeth
    Vishwavidyalaya
Indore, India

Rachana Kamble
Department of CSE
Technocrats Institute of Technology
    Excellence
Bhopal, India

Nishakar Kankalla
Dept. of ECE
St. Martin's Engineering College
Secunderabad, India

Saima Khan
Department of Electronics and
    Communication Engineering
Technocrats Institute of Technology
    (Excellence)
Bhopal, India

Sateesh Kourav
Department of Electronics and
    Communication Engineering
Indian Institute of Information
    Technology, Design and
    Manufacturing (IIITDM)
Jabalpur, India

Adarsh Mangal
Department of Mathematics
Engineering College Ajmer (An
    Autonomous institute of
    Government of Rajasthan)
Kiranipura, India

Amar Nayak
Department of CSE
Technocrats Institute of Technology
    Excellence
Bhopal, India

Tushar Kumar Pandey
Department of Computer Science
Dr. Rajendra Prasad Central
    Agricultural University
Pusa, Samastipur, Bihar, India

Geetam Richhariya
Department of Electrical &
    Electronics Engineering
Oriental Insititute of Science &
    Technology College
Bhopal, India

Vandana Roy
Department of EC
Gyan Ganga Institute of
    Technology and Sciences
Jabalpur, India

Shravan Kumar Sable
Department of Electronics &
    Communication
LNCT Bhopal
Bhopal, India

Manish Sawale
Department of Electrical &
    Electronics Engineering
Oriental Insititute of Science &
    Technology College
Bhopal, India

Devkant Sen
Department of Electronics and
    Communication engineering
Technocrats Institute of Technology
    Bhopal
Bhopal, India

KV Shahnaz
Department of ECE
Vel Tech Rangarajan Dr Sagunthala
    R&D Institute of Science and
    Technology
Chennai, India

Raghavendra Sharma
Department of Electronics &
    Communication Engineering
Amity University
Gwalior, India

T M Shashidhar
Department of CSE
Acharya Institute of Technology
Bengaluru, India

Dipti Shukla
Department of Physics
O.P Jindal University
Raigarh, India

Rajesh Kumar Shukla
Department of Computer
    Science
Oriental Insititue of Science &
    Technology College
Bhopal, India

Siddharth Shukla
Department of Electrical &
    Electronics Engineering
Technocrats Institute of Technology
    Bhopal
Bhopal, India

Nagendra Singh
Department of Electrical
    Engineering
Trinity College of Engineering &
    Technology
Karimnagar, India

Anita Soni
Department of Computer Science
    and Engineering
IES University, Bhopal, India

R Sreelakshmy
Department of ECE
Vel Tech Rangarajan Dr Sagunthala
    R&D Institute of Science and
    Technology
Chennai, India

M Sundararajan
Department of Mathematics and
    Computer Science
Mizoram University
Aizawl, India

Aby K Thomas
Department of ECE, Alliance
    College of Engineering and
    Design
Alliance University
Bengaluru, India

Rovin Tiwari
Department of Electronics &
    Communication Engineering
Amity University
Gwalior, India

Kirti Verma
Department of Engineering
    Mathematics
Gyan Ganga Institute of
    Technology and Sciences
Jabalpur, India

V Vishnupriya
Electronics Department
Government Model Engineering
    College
Kochi, India

Nita Vishwakarma
Department of Electrical &
    Electronics Engineering
Oriental Insititue of Science &
    Technology College
Bhopal, India

Chapter I

Antenna design for IoT and biomedical applications

Aby K Thomas
Alliance University, Bengaluru, India

Tushar Kumar Pandey
Dr. Rajendra Prasad Central Agricultural University, Pusa, Samastipur, India

Madhukar Dubey
ITM College, Gwalior, India

T M Shashidhar
Acharya Institute of Technology, Bengaluru, India

Vandana Roy
Gyan Ganga Institute of Technology and Sciences, Jabalpur, India

Nishakar Kankalla
St. Martin's Engineering College, Secunderabad, India

1.1 INTRODUCTION

Today's world is dominated by a culture in which interpersonal communication centered on the human body is the norm. Through its sensor nodes, antennas, and processing units, the Wireless Body Area Network (WBAN) is an essential component of body-centric communication. The WBAN can be used in conjunction with an external system, a system worn on the skin, or through theimplantation of a device directly into the human body. Therefore, it is commonly broken down into two subtypes: in-body and out-of-the-body [1].

The body area network antenna is a planar antenna that can be used for body-centric communication and has the advantages of being flexible, biocompatible, and lightweight. The substrate class requires a malleable substance. The conductive component must be comfortable and easy to wear.

Both the range and the utility of biomedical antenna technology are extremely high. Biotelemetry, biomedical treatment, and medical diagnosis

DOI: 10.1201/9781003422440-1

1

are just few of the fundamental uses. From the 1960s onward, we began to see the introduction of gadgets such as implantable pacemakers and pill computers. The biomedical industry has invested heavily in R&D to improve systems and expand their uses. The miniaturization of antennas, increased antenna efficiency, packaging with adequate insulation, and antenna performance optimization in lossy biological tissues are fundamental aspects of biomedical systems [2, 3].

Biotelemetry makes use of wireless medical equipment to collect patient information and transmit it to a central location. Patients can be monitored both remotely and in realtime with the aid of biotelemetry. Biomedical therapy is another field that uses implantable devices for disease diagnosis and treatment [4–6]. Biomedical devices can use one of two frequency bands designated by the International Telecommunication Union (ITU) for such purposes. Biomedical antennas have power constraints that must be met. They become detrimental to human health if they are not consumed properly [5]. As a result, the creation of effective biomedical antennas is no easy feat. The behavior and performance of these antennas are determined by the surrounding human body losses. Therefore, antennas need to be fine-tuned across a wide range of metrics, including size, performance, power consumption, etc.

Biomedical antennas are a useful component of remote health monitoring systems, as seen in Figure 1.1. Using this biotelemetry device, doctors and their careers may keep tabs on their patients' well-being. However, modern

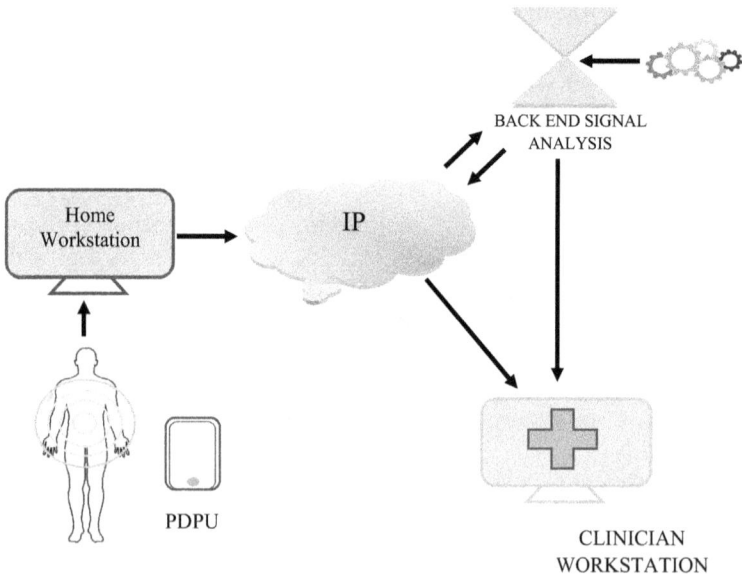

Figure 1.1 Remote health monitoring system.

technology calls for small, low-power, high-efficiency systems [6–8]. This motivates much study, ultimately leading to the recommendation of employing a number of antennas to satisfy the need.

Telemetry data collection, medical diagnosis, and medical therapy are three of the most important medical applications. Telemetry data is concerned with wireless information transfer between medical devices and central processing units. Data from a variety of biological sensors will be transmitted to the processing units [9]. Pulse rate, glucose level, body temperature, heart rate, and many more metrics can all be monitored with the use of existing biomedical sensors. The elderly, those with chronic diseases, athletes, and health fanatics can all benefit from data telemetry. Significant disorders, such as water retention, breast cancer, and heart attacks, can be detected with the help of the medical diagnosis technique [10]. For instance, before the advent of microwaves, the diagnosisof breast cancer was a painful process. The early stages of breast cancer can now be identified rapidly and painlessly by doctors using a portable gadget.

One technology that makes use of microwaves' heating properties is medical therapy. Microwaves can be used to kill cancer cells by heating up malignant tissues within the body. As a result, this treatment option is preferable to chemotherapy in terms of both efficacy and comfort [11]. X-rays were the first medical application of electromagnetic radiation. The first pill cameras for endoscopy were created in the 1960s. In addition, cardiovascular gadgets became a reality. Multiple devices have been created since that time, and scientists are always striving to perfect them. There is an increasing need for telemedicine services such as remote therapy and remote patient monitoring. There is a growing demand for increasingly sophisticated biomedical devices in the field of smart homes. Biomedical antennas need to be improved so that they can keep up with the demands of modern medicine. Antenna characteristics must be fine-tuned [12], depending on the intended use.

Patients can now receive treatment at home with the help of an increasingly popular wireless health monitoring device. Treatment efficiency is greatly aided by wireless communication linkages between doctors and patients [13]. Cheap price, cheap maintenance, easy wearability, and low power consumption are among the fundamental requirements for such systems. Sensors will be added to the system, and data will be communicated to doctors as needed, depending on the patient's condition [14].

There are three broad types of wireless body-centric communication. The first is inter-network communication that occurs outside the human body. The second type of communication is nonverbal. This kind of interaction takes place between the implanted devices themselves. The third type is known as "on-body communication," and it refers to interactions between wearable gadgets. Bandwidth is essential for these systems to function well.

1.2 MOTIVATION FOR THE RESEARCH WORK

- Wearable antennas are a crucial component of body-centric communication systems. The antenna's gain and directivity are impressive, and it may be used with either an on- or off-body network. The primary objective of this effort is, therefore, to improve the small antenna's directional performance.
- Since the antenna's strength may need to be increased for effective reception and transmission during on-body communication, the radiation may be outward during this kind of interaction. Because of the dangers posed by antenna radiation that is reflected back onto the user's body.
- This chapter's focus is on how to pick the most effective antennas for use in the medical field. Its aim is to find the current constraints and compromises of standard designs. You can redesign the classic antenna using a variety of up-to-date optimization methods andbuild and test virtual versions of the planned antenna. An analysis can be made of how well it works in a media centered on the human body and blueprints can be advanced for a wide range of biomedical uses.

1.3 EXISTING WORK DONE

Antennas for body-centric communication is a topic that has spawned a great deal of academic literature. In particular, metamaterial application in antenna design and development have lately gained traction. Negative permittivity and permeability are what make it special, and they will help with aspects such as antenna size reduction, bandwidth improvement, gain, and directivity. Metamaterial structure is also used to successfully reduce backward waves in the context of wearable body area networks [15].

Researchers came up with a T-shaped monopole antenna. T-shaped devices can operate in many frequency bands and feature circular polarization. Planar Inverted F Antenna (PIFA) has now been described after three years of study, and it may be found in mobile phones that use the 900 MHz to 1900 MHz frequency range [16].

The authors presented a compact octaband antenna for wireless networking. Using a metamaterial allowed for miniaturization, it also enabled multiband operation. Similarly, with an operating frequency of 3.51 GHz, researchers presented an L-shaped antenna with a modification to the ground construction. The antenna has a multiband resonance; however, it is challenging to place it in a mobile device case [17]. To achieve negative permeability, a double layer of symmetric SRRs (D-SSRRP) was placed on either side of the antenna.

The directivity was improved by narrowing the radiation beam as the number of cells increased. When operating at higher frequencies, the radiation pattern improved. When compared to a standard patch antenna, the antenna's half power beam width (HPBW) is narrower.

In a similar vein, the scientists introduced a broadband antenna with enhanced gain in a specific direction. The conformal antenna features an array of three fork-shaped components, and parasitic patches featuring three triangles have been incorporated on both sides of the substrate at a scale of three by six unit cells [18]. Therefore, the structure in that area was responsible for the increase in speed. Furthermore, the configuration operates as a lens antenna due to its limited beam width. The antenna's primary use is for WiMAX.

When it comes to analytical models, the straight superstrate dipole antenna is where most of the work is done. With the intention of improving the antenna's directivity, the authors presented a rapid cavity model for real-world microstrip antennas. Using the transmission analogy and the reciprocity theorem, we were able to solve the problem of the radiation field in the model of the cavity [19]. Here, SRRs were created for use on either side of the substratein which the dielectric material was characterized by its permeability and permittivity, two magnetic and electric properties. Full wave analysis only needed to take 2% of the time using this method.

There is a significant need for biomedical antennas in the healthcare and business sectors. It has been demonstrated that Planar Inverted F Antennas (PIFA) are effective biological antennas. These antennas have the ideal characteristics for use in biomedical devices implanted inside the human body since they resonate at the biomedical frequency range. If you're looking for an implantable antenna, PIFA is your best bet [20]. This antenna is small, has excellent near-field radiation performance, and can be customized to use two or even three frequencies.

The antenna's performance suffers because its resonance frequency drops when used in a body-centric medium. So, in order to solve the issue, we need more bandwidth. In addition, higher bandwidth improves performance in cancer detection applications. Therefore, enhancements to PIFA antenna bandwidth are necessary [21].

The loading method is an effective strategy for boosting bandwidth. This section presents the design principles upon which UWB PIFA is founded. The first technique involves adjusting the feed width and the distance between the feed and the shorting plate. The second strategy involves the addition of a parasitic element in the form of an inverted L along the feed plane. In the third approach, the feed and shorting post planes receive a parasitic rectangle element. All three methods are applied in tandem to traditional PIFA for the analysis [22].

The need for ultra-wideband spectrum is growing daily to accommodate medical and industrial biomedical equipment. Because of its small size and other advantageous properties, the Planar Inverted F Antenna (PIFA) is a great antenna design for use in the medical field. The PIFA loading method is used to increasebandwidth [23].

The human body is the primary medium for biomedical antennas. The body is a lossy system. The other characteristics are homogeneous and spread out. When an antenna is put on or inside a human body, its properties will change due to the influence of the medium [24].

Planar-geometry biomedical antennas are optimized for specific uses. The hexagonal patch antenna is more compact than its circular counterpart. As a result, we'll be looking at a hexagonal patch. In the biomedical frequency spectrum, a hexagonal patch with a conventional ground structure is designed, simulated, and manufactured. In order to shrink the overall footprint, conventional ground is swapped out for various faulty ground constructions.

Defective ground structures in the form of circular and hexagonal rings are superimposed on the regular ground. The conventional underground framework is digitally recreated. HFSS is used to model and examine the effects of damage to the ground's structures. The simulation outcomes of both the regular ground and the flawed ground structures are examined and contrasted. The results of these comparisons are used to improve the efficiency of antennas used in medical settings [25].

A faulty ground plane in an antenna prevents electricity from flowing freely. It functions as both a radiator and a part of the feeding mechanism. It modifies the antenna's effective input characteristics and emission pattern. The performance of antennas inside the human body can be studied with the help of accessible body models. Phantoms are a type of phantom. These models aid numerical study of human-introduced antennas.

When an antenna is placed inside a body-centric medium, its parameters change. The internal environment of a human being is extremely complex, heterogeneous, dispersive, and lossy. The developed antenna can be tested on a number of available body phantoms. The performance of an antenna can be studied by simulating it within a phantom of choice.

1.4 SOFTWARE TOOLS

Antennas are designed using simulation software to cut down on manufacturing costs. The antenna is the end result of the software being used to perfect the design of the device untill it meets the standard. All types of electromagnetic issues, from low to high frequencies, are resolved by using this microwave studio programme to simulate high-frequency components like antennas, filters, and couplers. The optical, thermal, and mechanical properties are the focus of this programme. Finite Integration in Technique (FIT) is the foundational methodology behind the programme.

1.5 THE PROPOSED METHOD AND RESULT ANALYSIS

A reactive near field exists in empty space. As a result, the power cannot be radiated or absorbed. However, in a lossy medium, the amount of power

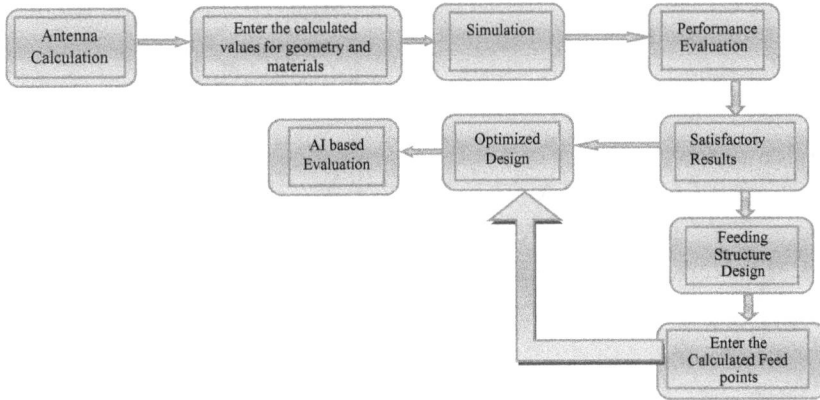

Figure 1.2 Simulation work flow.

lost rises due to near-field coupling with the nearest surrounding material. High near-field interaction with lossy tissues is responsible for the biomedical antennas' poor radiation efficiency. By employing insulation, the coupling issue can be mitigated. The radiation pattern is the elevation or azimuth map of electric and magnetic fields. The interference in a biomedical antenna's radiation pattern is caused by the human body.

Figure 1.2 presents the flow diagram for an effective antenna design that can be used for biomedical applications. Primary antenna computation parameters are decided, as are the antenna geometry and materials required for design, and then simulation is performed. Moreover, they were evaluated for satisfactory results. If, after testing, it is found correct, further application-oriented simulations are performed. Otherwise, check for feeding structure design and evaluate for optimised results through simulation. Finally, artificial intelligence-based analysis is performed to evaluate the effectiveness of biomedical and healthcare-based applications.

In general, impedance matching serves as a definition of bandwidth. The term "impedance bandwidth" can be used to describe this range of frequencies. Where the reflection coefficient is less than −10dB as a function of frequency. The bandwidth of a lossy medium is greater than that of a lossless medium. The electromagnetic field is strongly dissipated in the surroundings, and only a small fraction of the field is reflected back to the source when the radiator is positioned in a lossy medium. Bandwidth is shown to increase by a factor of 10 above what is seen in a lossless medium.

Due to power dissipation, the biomedical antenna becomes hot when put in the body-centric medium. The specific absorption rate (SAR) is a method for evaluating the interaction of electromagnetic radiation with living organisms. The dissipation of electromagnetic energy per mass unit is represented by SAR. When determining SAR, an average method is used, and a large number of peak values are taken into account. Biomedical antennas' maximum power dissipation must be in line with SAR regulations.

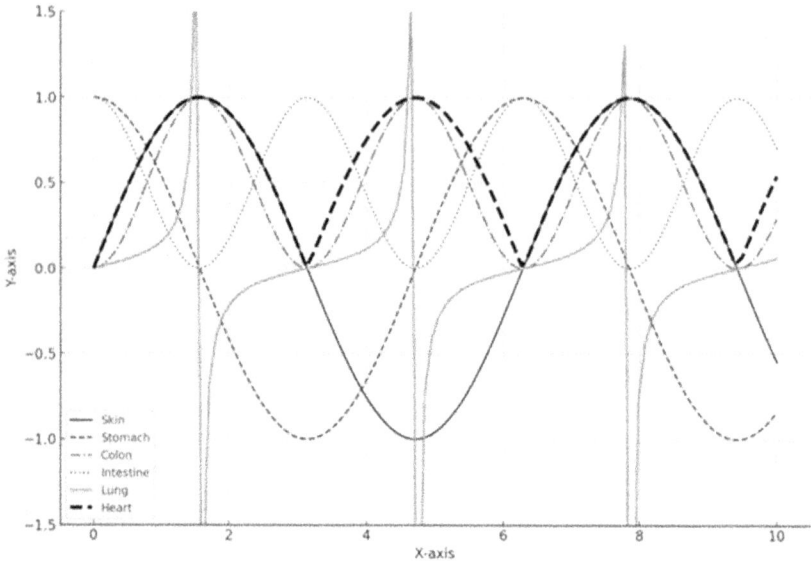

Figure 1.3 Variation of the resonance frequency of PIFA with different parts of the human body.

All HFSS-created antennas are put through their paces within a model human body for verification. This section displays the achieved outcomes. Three versions of PIFA with parasitic elements—a circular patch, a hexagonal patch, and a model of a PIFA with a defective ground are evaluated.

Figure 1.3 shows how the resonance frequency of PIFA varies with various human body regions. The skin medium moves the fundamental resonance frequency from 3 GHz down to 2.7 GHz. The antenna's signal strength is diminished due to malfunctioning components.

Resonance frequency shifts with various human body parts are depicted in Figure 1.4 for poorly grounded circular patches. Resonance occurs at a frequency of 1.98 GHz in a lossless medium. The resonance frequencies of the gastrointestinal tract, the colon, the epidermis, and the muscles all changed from 1.4 to 1.6 to 1.65 to 1.75 to 1.7 gigahertz.

Resonance frequencies change depending on where on the body they're being measured (see Table 1.1). The human body is a lossless medium with a dielectric constant of 1, which varies depending on the location. In a lossless medium, the resonance frequency for both defectively grounded circular and hexagonal patches is 1.93. PIFA operates at a resonance frequency of between 3 and 7.8 GHz. HFSS software is used to examine the resonance frequency shifts using five different body regions. Examinations of the intestines, stomach, colon, skin, and skeletal muscles are performed.

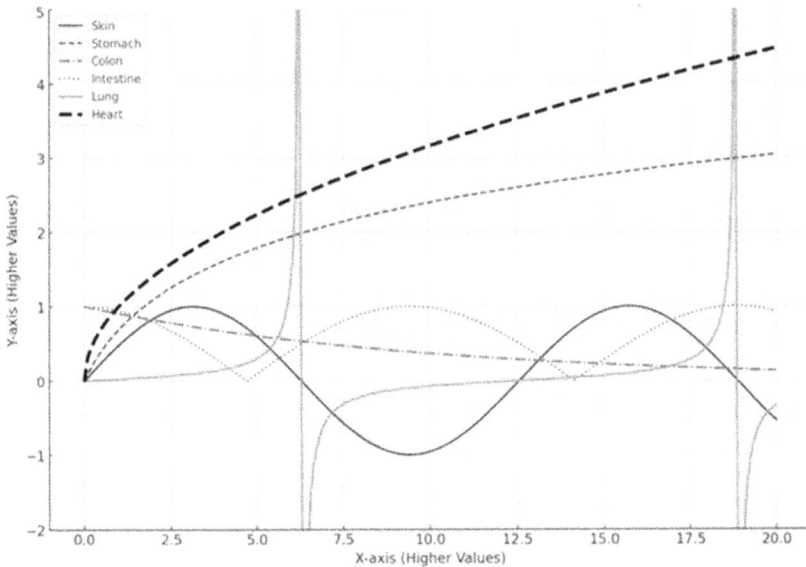

Figure 1.4 Variation of the resonance frequency of defective ground circular patch with different parts of human body.

Table 1.1 Resonance frequency variation for biomedical antennas with different parts of the human body

S. No.	Medium	Resonance frequency for defective ground circular patch	Resonance frequency for PIFA (GHz)
1	Lossless medium	1.93	3 to 7.8
2	Small intestine	1.62	2.8 to 10
3	Stomach	1.45	Not Reacting Appropriately
4	Colon	1.68	
5	Skin	1.79	
6	Muscle	1.72	

1.6 CONCLUSION AND FUTURE WORK

Merging biomedical gadgets have many potential medical and economic uses. Modern biomedical antennas that can match the needs of this developing technology are a must. There are two main types of biomedical uses: internal and external. Applications on the outside of the body include both therapeutic and commercial uses, whereas applications inside the body are mostly medical.

In this section, we use phantom models of the human body to analyze the performance of biomedical antennas that have been built. The human body's influence on the biomedical antennas' various parameters can be estimated with the use of this investigation. The 10dB bandwidth in the human body is significantly larger than the corresponding bandwidth in a lossless medium. In a lossy medium, the frequency of a resonance is lower than in a lossless one. Increasing bandwidth is one way to combat the frequency detuning impact produced by human body losses. The 10dB bandwidth in the human body is significantly larger than the corresponding bandwidth in a lossless medium. In a lossy medium, the frequency of a resonance is lower than in a lossless one. Increasing the operating bandwidth is one way to combat the frequency detuning impact brought on the human body losses.

However, additional investigation is required to address other concerns. As more and more uses for them emerge in the commercial and medical sectors, demand for biomedical devices continues to rise. To accommodate the needs of emerging technologies, biomedical antennas must be refined. Miniaturization, antenna performance, frequency detuning, biocompatibility, powering, packaging, manufacturing, and experimental characterization are some of the biggest hurdles for in-body applications.

There is a constant need for more efficient and compact biomedical antennas. It is planned to use wireless power transfer so that batteries will not need to be changed as frequently. Electromagnetic compatibility is an important factor for biomedical devices since they must be used in a body-centric environment. In the context of healthcare applications, patient confidentiality is of paramount importance. Inadvertent errors in wireless patient data transmission can expose the system to security risks. It is important to consider the system's particular absorption rate and any potential biological dangers when developing such a system. The body-centric system relies heavily on the functionality of the base station antenna.

REFERENCES

1. K.S. Nikita, *Handbook of Biomedical Telemetry*. John Wiley & Sons, Inc., Hoboken, New Jersey, 2014.
2. M. Ghovanloo Topsakal, and R. Bashirullah, "Guest editorial: IEEE AWPL special cluster on wireless power and data telemetry for medical applications," *IEEE Antennas and Wireless Propag. Lett.*, vol. 11, pp. 1638–1641, 2012.
3. A. Kiourti, and K.S. Nikita, "A review of implantable patch antennas for biomedical telemetry: Challenges and solutions," *IEEE Antennas and Opag. Magazine*, vol. 54, no. 3, pp. 210–228, Jun. 2012.
4. E.Y. Chow, M.M. Morris, and P.P. Irazoqui, "Implantable R.F. medical devices: The benefits of high-speed communication and much greater communication distances in biomedical applications," *IEEE Microw. Magazine*, vol. 14, no. 4, pp. 64–73, Jun. 2013.

5. L. Yang, C.L. Tsay, K.T. Cheng, and C.H. Chen, "Low-invasive implantable devices of low-power consumption using high-efficiency antennas for cloud health care," *IEEE J. Emerg. Sel. Top. Circuits Syst.*, vol. 2, no. 1, pp. 14–23, Mar. 2012.
6. A. Kiourti, and K.S. Nikita, "Design of implantable antennas for medical telemetry: Dependence upon operation frequency, tissue anatomy, and implantation Site," *Int. J. Monit. Surveillance Technol. Res.*, vol. 10, pp. 16–33, Jan–Mar. 2013.
7. K. Sajith, J. Gandhimohan, and T. Shanmuganantham, "SRR loaded CB-CPW fed hexagonal patch antenna for EEG monitoring applications," *2017 IEEE International Conference on Circuits and Systems (ICCS), Thiruvananthapuram,* 2017, pp. 66–70, DOI:10.1109/ICCS1.2017.8325964
8. H. Arshad Kahwaji, S. Sahran, A.G. Garba and R.I. Hussain, "Hexagonal microstrip antenna simulation for breast cancer detection," *2016 International Conference on Industrial Informatics and Computer Systems (CIICS)*, Sharjah, pp. 1–4, 2016, DOI: 10.1109/ICCSII.2016.7462400
9. H. Ullah, F.A. Tahir, A.S. Malik, and H.M. Cheema, "A Wearable Radiometric Antenna for Non-Invasive Brain Temperature Monitoring," *2018 18th International Symposium on Antenna Technology and Applied Electromagnetics (ANTEM)*, Waterloo, ON, pp. 1–6, 2018, DOI: 10.1109/ANTEM.2018.8572904
10. Y. Raza, M. Yousaf, N. Abbas, A. Akram, and Y. Amin, "A High Gain Low-profile Implantable Antenna for Medical Applications," In *Proceedings of the 2021 IEEE Asia Pacific Conference on Wireless and Mobile (APWiMob)*, Bandung, Indonesia, 8–10 April 2021, pp. 253–257.
11. N. Abbas, B. Manzoor, M. Yousaf, M. Zahid, Z. Bashir, A. Akram, and Y. Amin, "A Compact Wide Band Implantable Antenna for Biotelemetry," In *Proceedings of the 2019 Second International Conference on Latest trends in Electrical Engineering and Computing Technologies (INTELLECT)*, Karachi, Pakistan, 13–14 November 2019, pp. 1–5.
12. A. Lamkaddem, A.E. Yousfi, K.A. Abdalmalak, V.G. Posadas, L.E.G. Muñoz, and D. Segovia-Vargas, "A Compact Design for Dual-band Implantable Antenna Applications," In *Proceedings of the 2021 15th European Conference on Antennas and Propagation (EuCAP)*, Dusseldorf, Germany, 22–26 March 2021, pp. 1–3.
13. N. Pournoori, S. Ma, L. Sydänheimo, L. Ukkonen, T. Björninen, and Y. Rahmat-Samii, "Compact Dual-Band PIFA Based on a Slotted Radiator for Wireless Biomedical Implants" In *Proceedings of the 2019 IEEE International Symposium on Antennas and Propagation and USNC-URSI Radio Science Meeting*, Atlanta, GA, USA, 7–12 July 2019, pp. 13–14.
14. K. Fukunaga, S. Watanabe, and Y. Yamanaka, "Dielectric Properties of Tissue-Equivalent Liquids and Their Effects on Specific Absorption Rate" *IEEE Trans. Electromagn. Compat.*, vol. 46, pp. 126–129, 2004.
15. A.W. Damaj, H.M.E. Misilmani, and S.A. Chahine, "Implantable Antennas for Biomedical Applications: An Overview on Alternative Antenna Design Methods and Challenges" In *Proceedings of the 2018 International Conference on High Performance Computing & Simulation (HPCS)*, Orléans, France, 16–20 July 2018, pp. 31–37.
16. S.A. Hamzah, M.K. Raimi, N. Abdullah, and M.S. Zainal, "Design, Simulation, Fabrication and Measurement of a 900MHZ Koch Fractal Dipole Antenna," In *Proceedings of the 2006 4th Student Conference on Research and Development*, Shah Alam, Malaysia, 27–28 June 2006, pp. 1–4.

17. A. Akinola, G. Singh, and A. Ndjiongue, "Frequency-Domain Reconfigurable Antenna for COVID-19 Tracking," *Sens. Int.*, vol. 2, p. 100094, 2021.
18. N. Al-Fadhali, H.A. Majid, R. Omar, J. Abdul Sukor, M.K. Rahim, Z. Zainal Abidin, M. Al-Bukhaiti, A. M. Momin, and N. Abo Mosali, "Review on Frequency Reconfigurable Antenna Using Substrate-Integrated Waveguide for Cognitive Radio Application," *J. Electromagn. Waves. Appl.*, vol. 35, no. 7, pp. 958–990, 2021.
19. A.I. Al-Muttairy, and M.J. Farhan, "Frequency Reconfigurable Monopole Antenna with Harmonic Suppression for IoT Applications. *TELKOMNIKA (Telecommun Comput Electron Control)*, vol. 18, no. 1, pp. 10–18, 2020.
20. F.A. Asadallah, J. Costantine, Y. Tawk, L. Lizzi, F. Ferrero, and C.G. Christodoulou, "A Digitally Tuned Reconfigurable Patch Antenna for IoT Devices," In *2017 IEEE International Symposium on Antennas and Propagation & USNC/URSI National Radio Science Meeting*, IEEE, pp. 917–918, 2017.
21. J. Balcells, Y. Damgaci, B.A. Cetiner, J. Romeu, and L. Jofre, "Polarization Reconfigurable MEMS-CPW Antenna for mm-Wave Applications," In *Proceedings of the Fourth European Conference on Antennas and Propagation*, IEEE, pp. 1–5, 2010.
22. Z. Chen, H. Wong, J. Xiang, and S.Z. Liu, "Polarization-Reconfigurable Antennas for Internet of Things," In *2019 International Conference on Microwave and Millimeter Wave Technology (ICMMT)*, IEEE, pp. 1–3, 2019.
23. N. Singh and Y. Kumar, "Blockchain for 5G Healthcare Applications: Security and privacy solutions," *IET Digital Library*, vol. 1, pp. 1–30, 2021, DOI: 10.1049/PBHE035E
24. N. Singh, "IOT Enabled Hybrid Model with Learning Ability for E-Health Care Systems", *Measurement: Sensors*, vol. 24, pp. 1–14, Dec. 2022, DOI: 10.1016/j.measen.2022.100567

Chapter 2

Design of microstrip antenna for multipurpose wireless communication

Nagendra Singh
Trinity College of Engineering & Technology, Karimnagar, India

Priyamvada Chandel
Energy Meter Testing Laboratory Central Power Research Institute, Bhopal, India

Siddharth Shukla
Technocrats Institute of Technology, Bhopal, India

Anita Soni
IES University, Bhopal, India

Devkant Sen
Technocrats Institute of Technology, Bhopal, India

Saima Khan
Technocrats Institute of Technology (Excellence), Bhopal, India

2.1 INTRODUCTION

The massive amount of data used in wireless communication necessitates a high channel capacity in the spectrum. Several dual-polarized antennas with broadside radiation had to be proposed in earlier years, but varied spectrums needed everyday data utilization [1, 2]. Due to the cheap cost and cost-effectiveness of signal transmission at THz as well as the need for continuous broadband or wideband and high speed at a high data rate, which requires a THz antenna, the spectrum band for Terahertz (0.1 to 10 THz) has gained popularity over the past decade or so [3]. The transmission and reception of radio waves, which are crucial elements in wireless communication, are accomplished with the aid of an antenna [4]. Resonant antennas are those that operate successfully and efficiently in a specific frequency range. As a component of the system, the antenna functions as a transducer to convert electrical impulses into electromagnetic waves and vice versa [5].

DOI: 10.1201/9781003422440-2

A transmitting and receiving antenna ought to have similar polarization in order to achieve effective transmission. For high efficiency, the transmitting and receiving antennas ought to have matched internal impedances. The first and simplest category of antenna used was the dipole antenna, which is a straight wire fed from its centre. The length of the antenna is supposed to be half the wavelength of its operating frequency for the effective transmission and reception of electromagnetic waves [6]. For example, if the operating frequency (f_o) of an antenna is 300 MHz, then the wavelength (λ) of that antenna is calculated as follows:

$$\lambda = \frac{c}{f} \tag{2.1}$$

Where "c" represents speed of light in free space.

Wavelength is achieved as 1 m, and the required antenna height will be 0.5 m.

2.2 TYPES OF ANTENNA

As we previously discussed, an antenna is used for transmitting and receiving electromagnetic waves [7]. Antennas are classified on the basis of their configurations. There are primarily six types of antennas that are discussed below:

2.2.1 Wire antenna

As the name reveals, the wire antennas consist of wire of different shapes and sizes, such as a dipole that is a straight wire, a loop, and a helix, as shown in Figure 2.1 [8]. Loop antennas may be circular or can be of

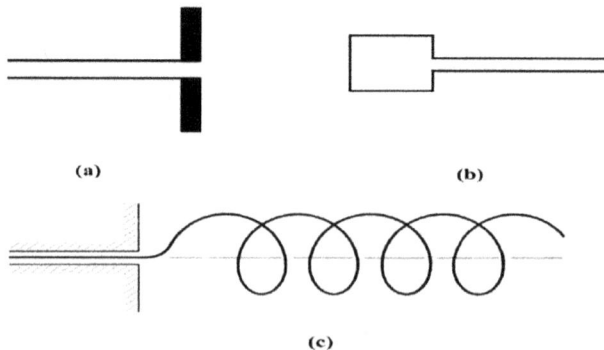

Figure 2.1 Configuration of wire antenna, (a) dipole, (b) square loop, (c) Helix.

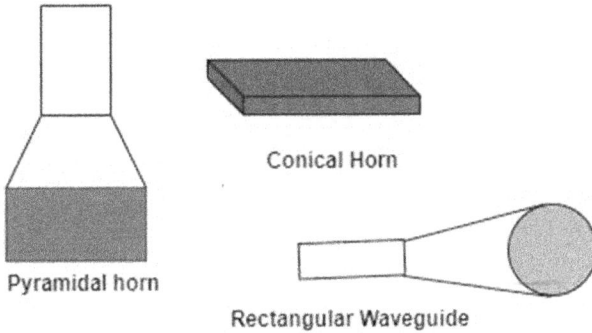

Figure 2.2 Configurations of aperture antenna.

any configuration, such as rectangle, square, or ellipse; however, circular loops are commonly used due to the ease of understanding and analysis. These are generally used in automobiles, spacecraft, ships, buildings, aircraft, etc.

2.2.2 Aperture antennas

These antennas are normally used with higher frequencies that are in microwave frequencies. There are several configurations of aperture antennas and some of them are generally used aperture antennas as shown in Figure 2.2 [9].

These antennas are used in aircraft and spacecraft applications.

2.2.3 Microstrip antennas

These antennas are normally applicable for both commercial and governmental purposes. Microstrip antennas consist of a ground plane on one side of the dielectric substrate and a metallic patch that is on the other side of the substrate, as shown in Figure 2.3. Patch antennas are small in size, lightweight, and conformable to nonplanar and planar surfaces, having a thin profile and therefore suitable for use in different fields [10].

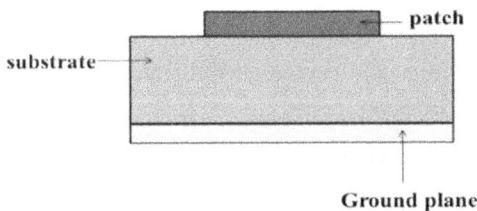

Figure 2.3 Rectangular microstrip patch antenna.

(a)

(b)

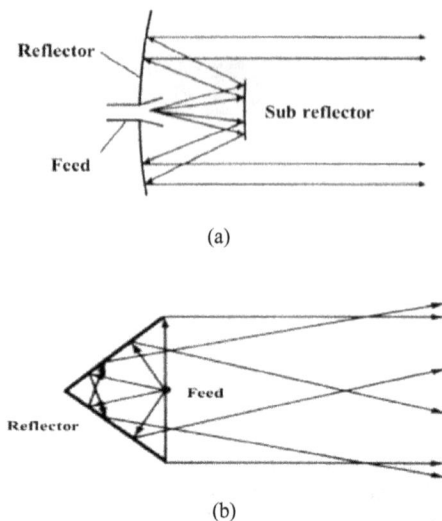

Figure 2.4 Reflector antenna, (a) parabolic reflector with cassegram feed, (b) corner reflection.

2.2.4 Reflector antenna

These antennas are used whenever communication over a long distance is required. In order to transfer signals over a long distance, say millions of miles, high-gain antennas are required. These types of antennas are intended for large diameters, as large as 300 m [11]. A parabolic reflector is one of the most frequently used antennas for such an application. Parabolic reflectors, corner reflectors, plane reflectors, and spherical reflectors are examples of reflector antennas and are shown in Figure 2.4. Parabolic reflectors are defined in two ways, depending on feed: front feed and cassegrain feed. Parabolic reflectors with front feed use only one reflector, which reflects the incident signal in the desired direction. Parabolic reflectors with cassegrain feed use a reflector and a sub-reflector; the incident wave is first reflected from the sub-reflector and then from the reflector in the desired direction. In the corner reflector case, if the angle made by the reflector is 90 degrees then it reflects the signal exactly in the same direction as it received [12].

2.2.5 Array antennas

When radiation characteristics are the main concern, array antennas are used. These are not made of a single element but are designed with an array of radiating elements. In order to maximize the radiation characteristics in a particular direction or desired direction and reducing them in other directions, an array of radiating elements is responsible. Some of the popular examples of array antennas are shown in Figure 2.5 [13].

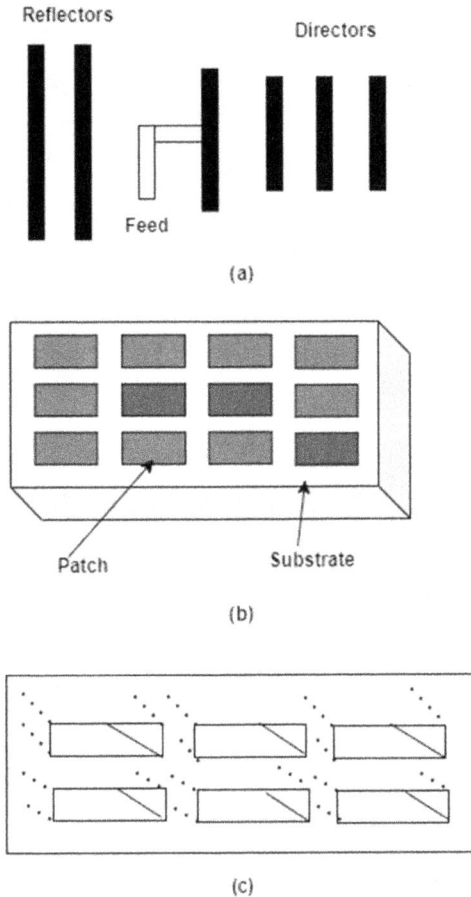

Reflectors

Directors

Feed

(a)

Patch   Substrate

(b)

(c)

Figure 2.5 Typical array antenna, (a) Yagi-Uda array, (b) microstrip patch array, (c) aperture array.

2.2.6 Lens antennas

Like parabolic reflectors, lens antennas are used at high frequencies and prevent the incident divergent energy from spreading in undesired directions. These antennas are classified on the basis of the material and geometry used to manufacture them. Some of the most typical examples of lens antennas are represented in Figure 2.6 with respect to the refractive index [14].

2.3 MICROSTRIP PATCH ANTENNA

Every day, the use of codeless communication systems increases, which raises interest in any efforts to enhance antenna performance. As was previously stated, microstrip antennas are more favorable than traditional microwave

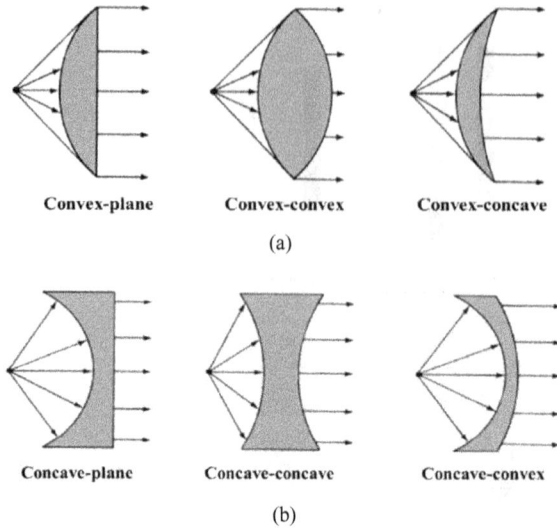

Convex-plane   Convex-convex   Convex-concave

(a)

Concave-plane   Concave-concave   Concave-convex

(b)

Figure 2.6 Typical example of the lens antenna, (a) example of lens antenna with refractive index (n) > 1, (b) example of lens antenna with refractive index (n) < 1.

antennas and are employed in a variety of applications, including radar, the global positioning system (GPS), satellite, and mobile communication, and telemedicine [15]. These patch antennas offer several benefits over traditional microwave antennas, including a narrow profile, lightweight, dual frequency and dual polarization, simple manufacture, etc. However, they are limited by several intrinsic drawbacks, such as a narrow impedance bandwidth of less than 5%, low gain, and poor efficiency. Using a dielectric substrate with a low dielectric constant and higher height can improve an antenna's performance in terms of efficiency, bandwidth, and radiation pattern. However, when the dielectric constant is smaller, the size of the antenna grows as the substrate's height rises. The price of the antenna size improves antenna performance. Over the years, a lot of study has been done to increase bandwidth enhancement while also improving other metrics with compactness. In order to reduce the size of the antenna and increase impedance bandwidth, certain approaches like shorted pins, shaped slots, post-gaps, or parasitic components can be utilized. However, each of these techniques has a unique side effect, high cross-polarization (XP).

2.3.1 Types of microstrip antennas

These antennas are extremely versatile. It may be designed in any shape or dimension. There are a large number of physical parameters for microstrip antennas compared with conventional microwave antennas. Patch antennas are classified into four different categories [16].

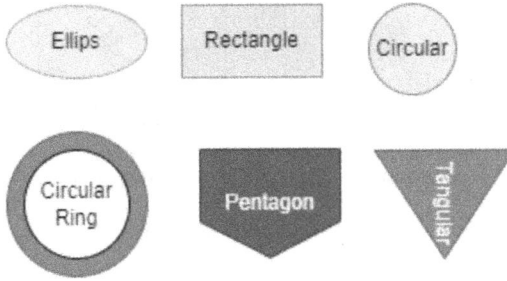

Figure 2.7 Commonly used shapes for patch antenna.

2.3.1.1 Microstrip patch antenna

These antennas are made up of a radiating patch on one side of the dielectric substrate and a ground plane on the other side. There are several shapes that are designed as patch antennas and classified as patch antennas. Some examples are shown in Figure 2.7. All of these configurations are of different shapes, but have the same radiation characteristics because they all behave like dipoles. Out of many configurations; circular patches and rectangular patches are generally used patterns due to the ease of analysis [17].

2.3.1.2 Printed dipole antenna

The length-to-width ratio of printed dipole antennas is different from that of microstrip antennas because the width of printed dipole antennas in free space is less than 0.0050. However, the radiation patterns of microstrip patch antennas and printed dipole antennas are similar, as both types of antenna have similar longitudinal current distributions, while cross-polar radiation, radiation resistance, and bandwidth are different for both antennas. These antennas have a small size and have linear polarization. And these antennas are well suited for higher frequencies with an electrically thick substrate and therefore can achieve significant bandwidth [18].

2.3.1.3 Microstrip or printed slot antenna

Printed slot antennas consist of a slotted ground plane. The slot can have any geometry that is integrated into the ground plane. The feeding technique used in slot antennas is microstrip feed, which is given by either a 50Ω microstrip feed line or a coplanar waveguide. Theoretically, the designs of microstrip patches may be designed as printed slots, as shown in Figure 2.8. These antennas are bidirectional in nature, and therefore they radiate in both directions, which minimize gain of the antenna [16].

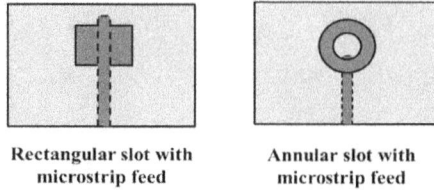

Rectangular slot with   Annular slot with
microstrip feed   microstrip feed

Figure 2.8 Printed slot antenna shapes with feed slot.

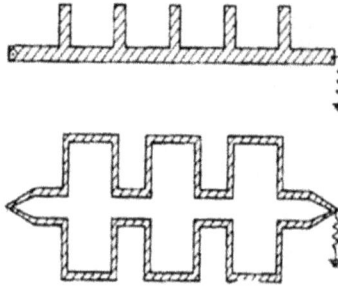

Figure 2.9 Configurations of microwave traveling wave antennas.

2.3.1.4 Microstrip traveling-wave antennas

Microwave traveling-wave antennas (MTA) support TE modes and consist of a chain of periodic conductors of adequate width or along a microstrip line. To avoid standing waves, the termination of the other end of this antenna with a matched resistive load is crucial. Configurations of microwave traveling-wave antennas are represented in Figure 2.9 [19].

2.4 FEEDING TECHNIQUES

The two primary kinds of feeding methods—contacting and non-contacting—are utilized to excite the antenna. These contacting and non-contacting categories are further divided as follows:

- Coaxial feed/Probe coupling
- Microstrip feed
- Proximity-coupled feed
- Aperture-coupled feed.

For designing an antenna, the key issue is choosing the feeding technique. The very significant thought is impedance matching for the efficient transmission of power among radiating patches and feeding structures. Some feeding techniques use two different substrates, which need proper alignment and, if not achieved, introduce discontinuities. Further discontinuities tend to produce surface wave loss and spurious radiation. Spurious radiation is an undesired radiation that causes side lobes to generate with the main lobe. However, side lobes are always present in every antenna's radiation pattern, and it is significant to reduce power levels in side lobes to improve radiation patterns. Spurious radiation also causes cross-polar amplitude in the radiation pattern. While designing an antenna, the feeding method is a significant factor to diminish the unwanted spurious radiation and power level in the side lobe of the radiation pattern [20].

2.4.1 Coaxial feed/probe coupling

It is a cheap, easy, and efficient way of feeding. In this feeding technique, the inner conductor of the coaxial line is coupled to the ground plane via an N-type connector, and a hole is drilled in the substrate from the centre of the coaxial and shouldered into the patch, as shown in Figure 2.10. However, for thick substrates ($h > 0.02\lambda0$), the drilling of holes makes the structure completely nonplanar. Also, for thicker substrates, probe length increases, which results in an input impedance that is extra inductive, something which tends to create impedance matching problems [21]. This technique is beneficial because the feed can be located in any place within the patch that matches its input impedance. However, it provides a narrow bandwidth.

2.4.2 Microstrip feed

This method is very effortless to design. We can simply analyze. The main shortfall of this technique is that it contributes spurious radiation. From Figure 2.11, we can observe that in this method, the feeding line is directly attached to the conducting patch of the microstrip antenna. Coupling of power takes place between the microstrip feed line and radiating patch and could be either edge-coupled or gap-coupled [22].

In edge coupling, the input impedance of the patch is very high at its radiating edge compared to 50Ω of the feed line, and therefore an external impedance matching circuit is necessary. The radiation through a part of the

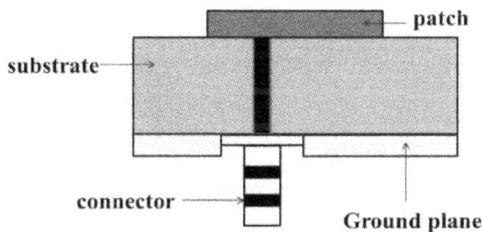

Figure 2.10 Probe feed rectangular microstrip antenna.

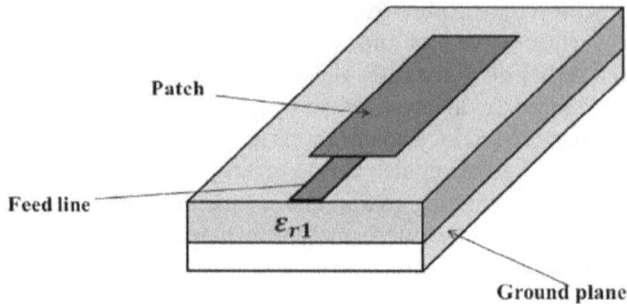

Figure 2.11 Microstrip feed in patch antenna.

patch that has contact with the feed line is blocked, thus reducing the radiation pattern. In gap coupling, proficient power coupling is probable with the help of a narrow gap. An antenna's power handling capability is reduced because of this narrow gap.

2.4.3 Proximity coupled feed

This is an example of a non-contacting, non-coplanar structure that is expressed in Figure 2.12. In this kind of feeding technique, two different types of substrate are used. The conducting patch is above the upper layer of substrate. The feed line is above the lower layer of the dielectric substrate. The microstrip feed line is acting as a sandwich among the different dielectric substrates [23]. This type of coupling that exists between the conducting patch and feed line is capacitive, and the power is coupled electromagnetically. This coupling is also known as an electromagnetic coupling. The feed line terminates with an open end, which is considered a stub and used for bandwidth improvement as the variation in length of the stub tends to cause disparity in the reflection coefficient and hence the return loss level.

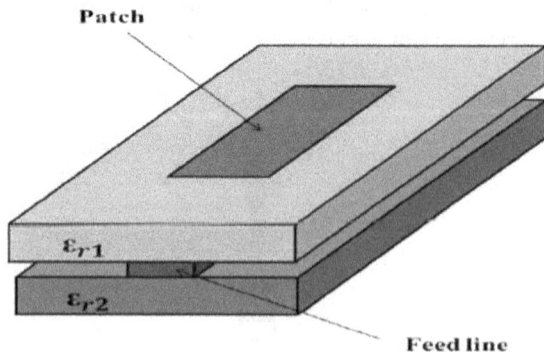

Figure 2.12 Proximity coupled feed.

Impedance matching plays a major role in antenna design. By varying the length of the feeding line, or stub, and using the appropriate dimensions of the patch, impedance matching can be obtained. Because it has two layers of dielectric, a suitable alignment is essential to avoid losses that make its fabrication difficult, which is considered the drawback of this technique. To decrease spurious radiation and improve impedance bandwidth, the lower layer has to be thin, and the upper layer has to be twice as thick as the lower layer. Two layers of substrate boost the overall thickness of the antenna, which extenuates the redundant spurious radiation and thus enhances the impedance bandwidth of the antenna (as high as 13%) [24].

2.4.4 Aperture coupled feed

This feed method uses two layers of dielectric substrate. The ground plane acts as a sandwich between the dielectric substrate shown in Figure 2.13. In the ground plane, a coupling aperture is used to couple the patch and feed line electromagnetically. The level of coupling is determined by the size, shape, and position of the coupling aperture in the ground plane. There is no limitation on the size and shape of the coupling aperture. The dimensions of the aperture slot are responsible for achieving a wider impedance bandwidth. The changing position of the slot tends to change the characteristics of the antenna [25].

Likewise, in a proximity-coupled feed, the lower substrate is supposed to be thin with a high dielectric constant, and the upper substrate should be thick with a low dielectric constant. The ground plane between two substrates works as a shield; therefore, from the open end of the feed line, the radiation does not interfere with the radiation pattern of the patch and

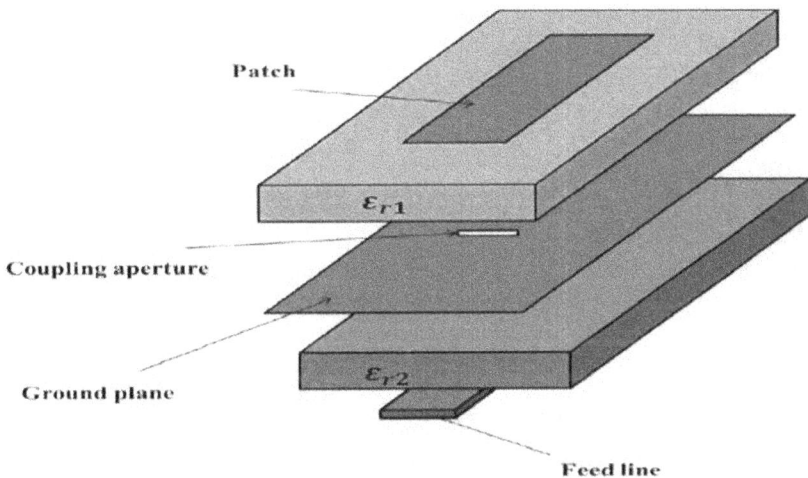

Figure 2.13 Aperture coupled microstrip feed.

therefore improves the polarization purity. To accomplish lower cross-polarization, the coupling aperture is generally designed in the center of the patch. A centered aperture slot confirms the proportion of configuration and hence causes lower cross-polarization. In this technique, the major drawback is that it is not simple to fabricate because of the multiple layers. Multiple layers also tend to increase the thickness of the antenna, and this method also gives a narrow bandwidth. In the contacting feeding technique, RF power is directly fed into the contacting patch using a contacting element like a connector, while proximity-coupled feed and aperture-coupled feed are types of non-contacting feeding methods in which RF power is transmitted via electromagnetic coupling.

2.5 DESIGNING AND RESULTS

2.5.1 Design specification

Using the concept of slot cutting, a patch antenna is proposed for C- and X-band applications. The corners of the exciting patch are truncated. An antenna is planned on a standard SiO_2 substrate at an operating frequency of THz. On the Integrated Electromagnetic 3 Dimension tool, simulations of the proposed antenna are performed and their attributes are assessed [26]. Follow these procedures to thoroughly analyze the structure:

- Formulation
- Designing
- Simulation and results

At first, the designing formulas are used to calculate the dimensions of the design structure together with the coaxial probe feed. It consists of calculating the dimensions of the ground plane and the probe feed. The following formulation is to be done for the design of the proposed structure: An inexpensive FR-4 substrate with a thickness of 1.5 mm is used for design purposes [27].

2.5.1.1 Formulation

It is a major step in the design of an antenna. It includes some important formulae to compute dimensions and other parameters that are essential for designing purposes.

2.5.1.2 Designing of ground plane

The geometry of an exiting patch with a parasitic element and shorting pin is shown in Figures 2.14 and 2.15, which show the top view of the micro strip antennas [28]. The dimensions, length (L_g) and width (W_g) of the

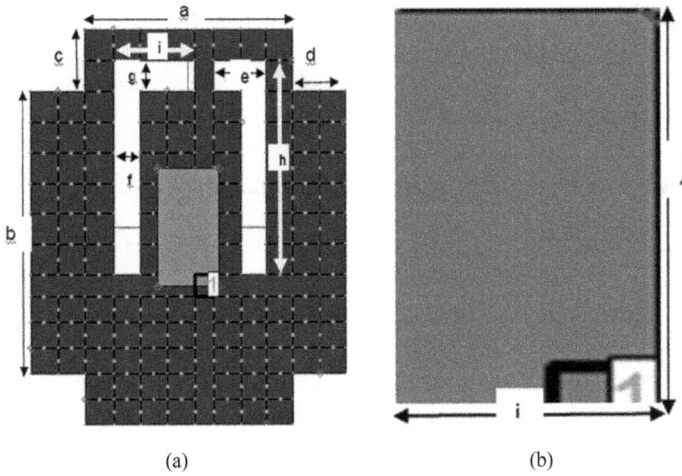

Figure 2.14 (a) and (b) Geometry of exiting patch with parasitic element with shorting pin.

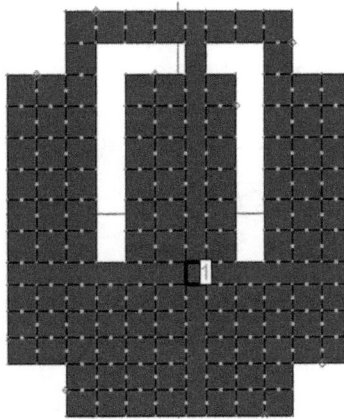

Figure 2.15 Top-view proposed geometry.

ground plane are calculated by using the dimensions of the patch antenna as described in equations 2.2 and 2.3:

$$L_g = L + 6h \tag{2.2}$$

$$W_g = L + 6h \tag{2.3}$$

In the above formulae "h" is height of substrate. The dimensions of the patch antenna are represented as follows: L represents length (in mm) and W represents width (in mm)

Microstrip patch antenna's width is formulated as,

$$W = \frac{1}{2fr\sqrt{\mu_0\varepsilon_0}}$$

(2.4)

Microstrip patch antenna's length is formulated as,

$$L = L_{eff} - 2\Delta L$$

(2.5)

Where L_{eff} is the effective length of patch and it is defined as

$$L_{eff} = \frac{c}{2f_0\sqrt{\varepsilon_{eff}}}$$

(2.6)

Where c is the speed of light. ΔL is the extended length of patch; it is defined as

$$\Delta L = 0.412h\frac{(\varepsilon_{eff} + 0.3)\left(\frac{w}{h} + 0.2664\right)}{(\varepsilon_{eff} - 0.258)\left(\frac{w}{h} + 0.8\right)}$$

(2.7)

$$\varepsilon_{eff} = \frac{\varepsilon_r + 1}{2} + \frac{\varepsilon_r - 1}{2}\sqrt{\frac{1}{1 + 12\frac{h}{w}}}$$

(2.8)

Where ε_{eff} is the effective value of dielectric constant of substrate.

2.5.1.3 Feeding points

Feeding point of the micro strip antenna is calculated as

$$X_f = \frac{L}{2\sqrt{\varepsilon_{eff}}}$$

(2.9)

$$Y_f = \frac{W}{2}$$

(2.10)

2.5.1.4 Design considerations

The proposed structure is simulated and analyzed on the Integrated Electromagnetic 3 Dimension tool. The study of the planned antenna depends on

Table 2.1 Mathematical calculation of the proposed antenna

Parameters	(mm)	Parameters	(mm)
a	8	g	1
b	10	h	7
c	2	i	2.5
d	2	j	3.8
e	2		
f	1		

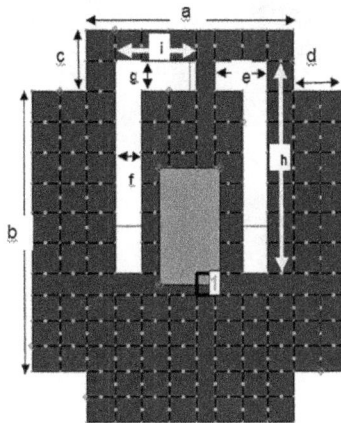

Figure 2.16 Geometry of planned antenna.

the effective values of the Si substrate. A substrate having a height of 1.5 mm is considered. The ideal value of the dielectric constant of the substrate is 11.9. The ideal value of the loss tangent of a SiO_2 substrate is 0.019. A ground plane of dimensions $10 \times 8 \times 1.6$ mm^3 is used to fabricate the prototype. Table 2.1 shows the parameter values used in calculations [29].

Figure 2.16 shows the proposed design and a cutting slot is used to integrate on microstrip patch. The exciting patches are U shape slotted for impedance matching.

Figure 2.16 depicts the exciting patch antenna's geometry in relation to the ground plane. The shorting pin is linked between the stimulated patch and ground, as illustrated in Figure 2.16, and the suggested antenna consists of a loaded 12-square rectangular slot in the ground layout. The H and U shape slots were employed in the earlier work, which offered a higher bandwidth boost than prior efforts. A shorting pin can be connected through a hole between the ground and the stimulated patch to further increase the proposed antenna's bandwidth. Figure 2.17 displays a 3D image of the suggested antenna with probe feeding.

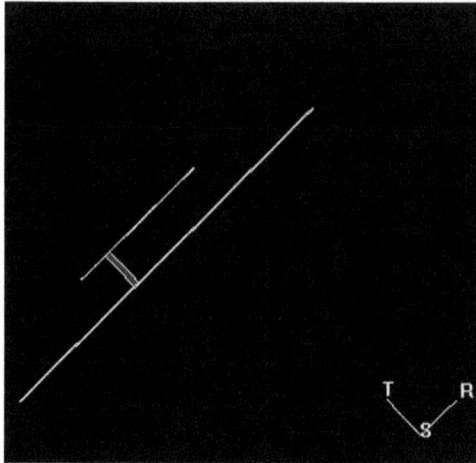

Figure 2.17 3D view of proposed structure.

2.6 CONCLUSION

For THz applications, a microstrip patch antenna with a THz frequency band is provided. The findings for the bandwidth and return loss are improved when the slot is introduced during an exciting time. The silicon dioxide substrate for the antenna is used, and IE3D software is used for simulation. Additionally, the VSWR and radiation efficiency are attained within a reasonable range. The suggested antenna's parameters have all been determined for the THz frequency range. The suggested antenna produced an acceptable radiation pattern within the frequency range and a band of around 55%. SiO_2 is used as the substrate for the antenna because of its low profile, low cost, and thickness. The probe feeding method is employed.

REFERENCES

1. Libin Sun, Yue Li, and Zhijun Zhang, "Wideband Dual Polarized Endfire Antenna Based on Compact Open Ended Cavity for 5G Mm Wave Mobile Phones," *IEEE Transaction on Antenna and Propagation*, vol. 70, no. 3, pp. 1632–1642, Mar. 2022.
2. Inzamam Ahmad, Sadiq Ullah, Adnan Ghaffar, Mohammad Alibakh Shikenari, Salahuddin Khan, and Ernesto Limiti, "Design and Analysis of a Photonic Crystal Based Planar Antenna for THz Applications," *Electronics*, vol. 10 2021, DOI: 10.3390/electronics10161941
3. Neng-Wu, Zhong-Xun Liu, Zhi-Ya Zhang, and Guang Fu, "Frequency-Ratio Reduction of a Low Profile Dual Circular Polarized Patch Antenna Under Triple-Resonance," *IEEE Antennas and Wireless Propagation Letters*, vol. 19, no. 10, pp. 1689–1693, Oct. 2020.

4. Sai Radavaram, Maria Pour, "Wideband Radiation Reconfigurable Microstrip Patch Antenna Loaded With Two Inverted U Slot," *IEEE Transactions on Antenna and Propogation*, vol. 67, no. 3, pp. 1501–1508, Mar. 2019.
5. Joysmita Chatterjee, Akhilesh Mohan, and Vivek Dixit," Broadband Circular Polarized H-Shaped Patch Antenna Using Reactive Impedance Surface," *IEEE Antenna and Wireless Propagation Letters*, vol. 17, no. 4, pp. 625–628, Apr. 2018.
6. Hang Wong, Kwok Kan So, and Xia Gao, "Bandwidth Enhancement of a Monopolar Patch Antenna with V-Shaped Slot for Car to Car and WLAN Communication," *IEEE Transaction on Vehicular Technology*, vol. 2016, pp. 01–07, 2016.
7. J.-S. Hong, E.P. McErlean, and B. Karyamapudi, "Eighteen-Pole Superconducting CQ Filter for Future Wireless Applications," *IEE Proc Microwave Antennas Propagate*, vol. 1, no. 53, pp. 205–211, 2006.
8. KuKhandelwal Mukesh, Santanu Dwari, B.K. Kanaujia, and Sachin Kumar, "Design and Analysis of Microstrip Patch Antenna with Enhanced for Ku Band Applications," ELSEVIER, AEUE-51203.
9. C. Li, Q. Zhang, Q. Meng, L. Sun, J. Huang, Y. Wang, X. Zhang, A. He, H. Li, Y. He, and S. Luo, "A High-Performance Ultra-Narrow Bandpass HTS Filter and its Application in aWind-Profiler Radar System," *Supercond. Sci. Technol.*, vol. 19, pp. S398–S402, 2006.
10. Mahrukh Khan, "Characteristics Mode Analysis of a Class of Empirical Design Techniquesfor Probe-Fed, U-slot Microstrip Patch Antennas," *IEEE Trans. Antennas Propag.*, vol. 64, 2016, DOI: 10.1109_TAP 2016.2556705
11. Steven Weigand, Greg H. Huff, Kankan H. Pan, and Jennifer T. Bernhard, "Analysis and Design of Broad Band Single-Layer Rectangular U-Slot Microstrip Patch Antenna," *IEEE Trans. Antennas Propag.*, vol. 51, no. 3, Mar. 2003.
12. Y. Qian, et al., "A Microstrip Patch Antenna Using Novel Photonic Bandgap Structures," *Microwave J.*, vol. 42, pp. 66–76, Jan. 1992.
13. C.A. Balanis, "*Antenna Theory, Analysis and Design*," John Wiley & Sons, New York, 1997.
14. G.F. Khodaei, J. Nourinia, and C. Ghobadi, "A Practical Miniaturized U-Slot Patchantenna with Enhanced Bandwidth," *Progress in Electromagnetics Research B*, vol. 3, pp. 47–62, 2008.
15. M. Koohestani, and M. Golpour, "U-Shaped Microstrip Patch Antenna with Novel Parasitictuning Stubs for Ulta Wideband Applications," *IET Microw. Antennas Propag.*, vol. 4, no. 7, pp. 938–946, 2010.
16. Y.F. Cao "A Multi Band Slot Antenna for GPS/WiMAX/WLAN Systems," *IEEE Transactions on Antennas and Propagation*, vol. 63, no. 3, 2015, pp. 952–958.
17. Sandra Costanzo, and Antonio Costanzo, "Modified U-Slot Patch Antenna with Reduced Cross-Polarization," *IEEE Antennas Propag. Mag.*, vol. 57, no. 3, pp. 71–80, Jun. 2015.
18. J.A. Ansari, and B.R. Ram, "Analysis of Broad-Band U-Slot Microstrip Patch Antenna," *Microwave and Optical Technology Letters*, vol. 50, no. 4, pp. 1069–1073, Apr. 2008.
19. U. Chakraborty, "Compact Dual–Band Microstrip Antenna for IEEE 802.11a WLAN Application," *IEEE Antennas and Wireless Propagations Letters*, vol. 13, p. 407, 2014.

20. Emmi K. Kaivanto, Markus Berg, Erkki Salonen, and Peter de Maagt, "Wearable Circularly Polarized Antenna for Personal Satellite Communication and Navigation," *IEEE Transactions on Antennas and Propagation*, vol. 59, no.12, Dec. 2011.

21. A.K. Shackelford, K.F. Lee, and K.M. Luk, "Design of small size wideband microstrip-patch antennas," *IEEE Antennas Propag. Mag.*, vol. 45, no. 1, pp. 75–83, Feb. 2003.

22. A.A. Deshmukh, and G. Kumar, "Half U-slot loaded rectangular microstrip antenna," In *IEEE AP-S Int. Symp. USNC/CNC/URSINational Radio Science Meeting*, vol. 2, pp. 876–879, 2003.

23. Tayeb A. Denidni, and Larbi Talbi, "High Gain Microstrip Antenna Design for Broadband Wireless Applications," *Int. J. RF Microw. Comput.-Aided Eng.*, vol. 13, pp. 511–517, 2003.

24. C.S. Lee, and V. Nalbandian, "Impedance Matching of Dual-Frequency Microstrip Antenna with an Air Gap," *IEEE Trans. Antennas Propag.*, vol. 41, no. 5, pp. 680–682, 1993.

25. A.F. Sheta, A. Mohra, and S.F. Mahmoud, "Multi-Band Operation of Compact H-Shaped Microstrip Patch Antenna," *Microw. Opt. Technol. Lett.*, vol. 35, no. 5, pp. 363–367, Dec. 2002.

26. K.F. Tong, K.M. Luk, and K.F. Lee, "Wideband II-Shaped Aperture-Coupled U-Slot Patch Antenna," *Microw. Opt. Technol. Lett.*, vol. 28, pp. 70–72, Jan. 2001.

27. A. Shackelford, K.F. Lee, D. Chatterjee, Y.X. Guo, K.M. Luk, and R. Chair, "Smallsize Wide Bandwidth Microstrip Patch Antennas," In *IEEE Antennas and Propagation International Symposium*, vol. 1, Boston, MA, pp. 86–89, July 2001.

28. W.X. Zhang, C.S. Pyo, S.I. Jeon, S.P. Lee, and N.H. Myung, "A New Type of Wideband Slot-Fed U-Slotted Patch Antenna," *Microwave and Optical Technology Letters*, vol. 22, no. 6, pp. 378–381, Sep. 1999.

29. B.L. Ooi, E.S. Siah, and P.S. Looi, "A Novel Microstrip-Fed Slot-Coupled Self-Complementary Patch Antenna," *Microw. Opt. Technol. Lett.*, vol. 23, no. 5, pp. 284–289, Dec. 1999.

Chapter 3

Analysis and simulation of standard gain 18–40 GHz frequency band horn antenna

Kirti Verma

Gyan Ganga Institute of Technology and Sciences, Jabalpur, India

Sateesh Kourav

Indian Institute of Information Technology, Design and Manufacturing, (IIITDM), Jabalpur, India

M Sundararajan

Mizoram University, Aizawl, India

Adarsh Mangal

Engineering College Ajmer (An Autonomous institute of Government of Rajasthan), Kiranipura, India

3.1 INTRODUCTION OF ANTENNAS

3.1.1 The systemic communications

- Considerations like data integrity and authentication are crucial. Security problems are more likely to arise with wireless transmission systems.
- Antennas are used in wireless transmission systems to carry the signal from transmitter to receiver.
- Both analogue and digital communication systems require antennas.

A gearbox wire begins to radiate when it is energized by an alternating signal and flared at its open ends. Electromagnetic radiation is produced as a result of accelerated electron velocity. As a result, an antenna can be thought of as a transducer that transforms wire current into a Radio Frequency (RF) signal in an outdoor setting. Reciprocal antennas keep their conductivity constant whether transmitting or receiving. Antennas are generally resonant and have a rather restricted operating frequency range [8, 11]. Two conical conducting components that are positioned end to end make up a biconical antenna. Due to its shape, it is highly suited for both transmitting and receiving electromagnetic waves, and is frequently utilized for broadband

DOI: 10.1201/9781003422440-3

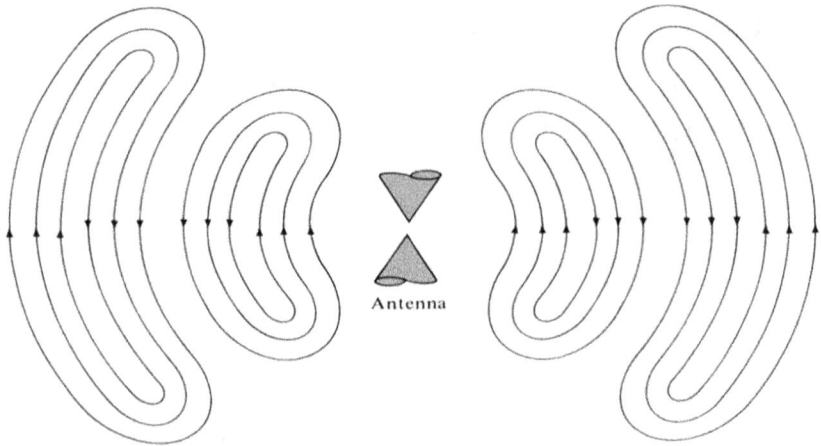

Figure 3.1 Bucolical antenna electric field lines for free-space waves.

applications, as shown in Figure 3.1. When a signal is applied to an antenna, it emits radiation in a certain pattern that travels over space. A radiation pattern visually displays how electricity spreads in empty space. There are many different types of antennas available today, and each one has its own unique characteristics and advantages. Therefore you can choose the antenna type that is best for the task at hand. The following is a list of the key antenna types.

- Because of its reciprocal nature, an antenna can function as both a transmitter and a receiver.
- Antennas exist in many different shapes and sizes, including printed, wire, and aperture antennas.
- Horn and slot antennas are examples of aperture antennas, whereas dipole, loop, and helix antennas are examples of wire antennas.

The two types of printed antennas are patch antennas and printed slot antennas [2].

This type of antenna offers certain advantages, such as portability and low cost, but it also has some drawbacks, such as restricted bandwidth.

3.2 THE HISTORY OF ANTENNAS

One of the main objectives of antenna research since Guglielmo Marconi developed the first wireless communication system in 1895 has been to find ways to expand the amount of information bandwidth that an antenna can use while also maintaining high radiation efficiency and reducing its physical size at any desired resonant frequency. When creating an antenna, compromises between these desired qualities are frequently required [1, 2]. Low-profile antennas may be necessary in high-performance military applications, such as those required for satellites, aircraft, spacecraft, and missiles, where cost, weight, size, performance, ease of installation, and aerodynamic geometry are limitations. These prerequisites are met by microstrip antennas, which explain why they have become so popular. The idea of a microstrip antenna that could operate simultaneously in the US and France was created around three decades ago by Decamps, Glutton, and Bassinet. Lewis understood immediately that strips line discontinuities are the source of radiation. In the second half of the 1960s, Kaolin conducted additional in-depth research on fundamental rectangle and square configurations. However, until Byron detailed a conducting strip line that was isolated from a ground plane by a dielectric slab in the early 1970s, no additional work, aside from the first Decamps investigation, had been documented in any publication. To power this antenna, which had a width of nearly half a wavelength and a length spanning many wavelengths, coaxial connectors were installed at regular intervals along both radiating sides. Munson quickly applied for a patent for a microstrip antenna [3]. Single-feed circularly polarized antennas are gaining popularity. Circular polarization is favourable for current and future commercial and military applications since the electric field vector is not required at the transmitting and receiving sites [4]. Wireless and mobile radio communications have equivalent corporate and government applications. When it is difficult to fit numerous orthogonal feeds with a power-splitting network, a single feed reduces the complexity, weight, and RF loss of the array feed. Circularly polarized microstrip antennas are small and light, conformal mountable, and compatible with microwave and millimeter wave integrated circuits (MMIC) [5]. The narrow bandwidth may be beneficial in some cases, such as secure communication systems.

3.2.1 Definition of antenna

In the broadcasting of radio waves into space, an antenna transforms them into alternating electromagnetic waves. Polarization is a step in the antenna selection and installation process. It is capable of both broadcasting and receiving signals. A metal structure that transmits or receives electromagnetic radiation is known as an antenna. In contrast, the antenna links the directing device into open space. It transmits a signal from the generator to space via transmission lines.

3.2.2 Radiation pattern

The radiation pattern of an antenna is one of the most essential factors impacting system coverage and performance. An antenna's radiation pattern determines how the energy it creates or receives is focused or directed [9]. Antennas typically do not emit more energy than is supplied to their input connection. Polar plots showing 360-degree angular patterns on vertical or horizontal sweep planes are frequently used to depict antenna radiation patterns. Decibels of relative power (dB) are used to measure it. The half power spots develop in the core of the lobe when the field intensity lowers to 0.707 of the maximum voltage. Radiation properties such as side lobes, rear lobes, and the front to back ratio are all of importance. It is not possible to completely remove the antenna's side and rear lobes. The impact of antenna side and rear lobes on antenna performance are many [6]. Side and back lobe energy originates in directions other than the designated coverage zone, resulting in waste. Energy delivered to the side and rear lobes of the transmitter may cause interference with other receiving devices. Many transmit sites' energy may be received at the receiver via the side and back lobes, causing system interference. Equation (3.1) is the radiation pattern of an isotropic array can be quantitatively modelled. -

$$F_E = \cos\cos\left(\frac{\beta d \cos\theta + \delta}{2}\right) \tag{3.1}$$

3.2.3 Gain of antenna

An antenna is a type of transducer that converts radio waves into alternating electromagnetic waves before transmitting them into space. Polarization is a phase in the selection and installation of an antenna. The vast majority of wireless communication systems use either linear or circular polarization. The directions of electric and magnetic field vectors of a plane electromagnetic wave are along positive y-directions and positive z-directions, respectively. The term "dB" refers to an isotropic radiator's gain. A half-wavelength dipole was employed to determine gain, which had an isotropic gain of 2.16 dB [7] Equation (3.2) reveals that the antenna's directivity is.

Directivity-

$$D = \frac{4\pi U}{P_{rad}} \tag{3.2}$$

U = radiation intensity
P_{rad} = total radiated power

$$\eta = \frac{P_{rad}}{P_{rad} + P_{lost}} \tag{3.3}$$

$$G = \eta D \tag{3.4}$$

3.2.4 Polarization

3.2.4.1 Gain of antenna - polarization

This is the spatial orientation of an electromagnetic wave supplied by a gearbox system's E-Field component. Low-frequency antennas are often vertically polarized due to ground reflection and the ease of physical manufacture, but high-frequency antennas are typically horizontally polarized [8]. Understanding polarization fluctuations is essential for the optimization of system performance. The antenna is a transducer that converts radio frequency into alternating electromagnetic waves before sending them into space [12]. Polarization is a phase in the selection and installation of an antenna. Most wireless transmission systems use either linear or circular polarization. The instantaneous field in the negative z direction of a plane wave transmission system. Equation 3.5 is a cross-product between an electric field that is x-polarized and a magnetic field that is y-polarized determines the direction of propagation, which is along the positive z-axis. The electric field is oriented along the x-axis; the magnetic field is directed along the y-axis.

$$E(z,t) = a_y E_x(z,t) + a_y E_y(z,t)$$ (3.5)

This is achieved by connecting the immediate components to their complicated counter parts. Equation 3.6 shows that a linearly polarized (LP) wave is one that has an electric field that points in the x direction while moving in the z direction. For someone whose electric field is polarized in the y direction, the same is true.

$$E_x(z,t) = E_{x0} \cos(\omega t + kz + \varphi_x)$$ (3.6)

$$E_y(z,t) = E_{y0} \cos(\omega t + kz + \varphi_y)$$ (3.7)

Where E_{x0} and E_{y0} are the magnitudes of the x and y components, respectively, and where the wave amplitude Eye is a constant and the factor k is related to frequency, as shown below. When an antenna is perpendicular to the Earth's surface, its electric field is vertically linearly polarized. Vertical polarization is illustrated via a vertical antenna atop a broadcast tower for AM radio or a whip antenna on an automobile. Horizontally linearly polarized antennas have a parallel electric field component to the Earth's surface. Horizontal polarization is used in television transmissions all around the world, for example [14]. As a result, television antennas are horizontally oriented. There must be a time-phase difference between two components for a wave to be linearly polarized.

$$\Delta\varphi = \varphi_y - \varphi_x = n\pi$$ (3.8)

where n = 0, 1, 2.

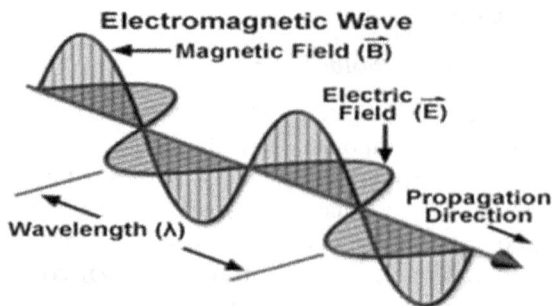

Figure 3.2 Transverse electromagnetic waves.

In circular polarization, the plane of polarization loops in a corkscrew pattern, completing one full cycle for each wavelength. In a circularly polarized wave, as seen in Figure 3.2, the properties of electromagnetic waves and mechanical waves creates energy in all planes, horizontal, vertical, and in between. Right-hand circular (RHC) rotation is used to describe clockwise propagation rotation. The left-hand circular (LHC) circles in an anticlockwise direction. Circular polarization can occur only when the magnitudes of two components are similar and their time-phase differences are odd [12]. Electromagnetic waves are transverse waves. The electric and magnetic fields fluctuate perpendicular to the propagation path of the wave. It's also worth noting that an EM wave's electric and magnetic fields are perpendicular to each other. Transverse electromagnetic waves have E and H fields that are perpendicular to the propagation direction. Any ideal conductor cross-section may be traversed by TEM waves. The y-axis represents the propagation direction. When the conductors on the left are energized by a source, they can provide electrical energy to a load connected between the right margins of the plates. In the future, they will be used as parallel plate transmission lines.

3.3 EFFECTS OF POLARIZATION

These directional fields have an effect on how an antenna transmits and receives radiation. A planar EM wave's electric and magnetic fields flow down a single channel. The polarization of an antenna is defined as the direction of electromagnetic fields formed by radiating energy. The two fields are perpendicular to one other and to the plane wave's path [10].

3.3.1 Reflectivity

The radio waves reflect as a result of their interaction with the material. When the reflecting surface does not totally reflect the signal in the same plane, the signal intensity is diminished, which is how linear polarized antennas solve concerns in one plane. A circularly polarized antenna broadcasts and receives in all planes. High-frequency devices that utilize linear polarization require two places to function properly [17]. Circularly polarized devices, on the other hand, receive reflected signals but return them in the opposite direction, preventing interference with the propagating signal. Circularly polarized signals may thus penetrate and bend around obstructions considerably more efficiently.

3.3.2 Multi-path

A main signal and a reflected signal arrive simultaneously at a receiver. This causes a "out of phase" situation. To detect, filter out, and analyse the proper signal, the radio receiver must use resources, reducing performance and speed. Because of the greater chance of reflection, linear polarized antennas are more prone to multipath.

3.3.3 Antennas with high gain and bandwidth

Antennas that must function effectively across a wide frequency range have been developed with the development of broadband wireless communication systems. Broadband antennas have an extremely wide bandwidth [7, 18]. However, how much bandwidth does a huge bandwidth the term "broadband" refers to a relative bandwidth measurement that depends on the situation. There are two ways to calculate bandwidth.

$$B = \frac{f_u - f_l}{f_c} \times 100\% \tag{3.9}$$

Where f_u and f_l are the highest and lowest operating frequencies at which adequate performance is achieved. f_c denotes the central frequency

$$B = \frac{f_u}{f_l} \tag{3.10}$$

A broadband antenna is a very variable notion that varies depending on the antenna. A broadband antenna is defined as one whose impendence and pattern do not change significantly over an octave $\left(\frac{f_u}{f_l} = 2 \right)$. Broadband antennas often require smooth edges a main signal and a reflected signal arrive at the same time to a receiver. This causes a "out of phase" situation. To detect, filter out, and analyse the proper signal, the radio receiver must use resources, reducing both performance and speed. Because of the greater chance of reflection, linear polarized antennas are more prone to multipath.

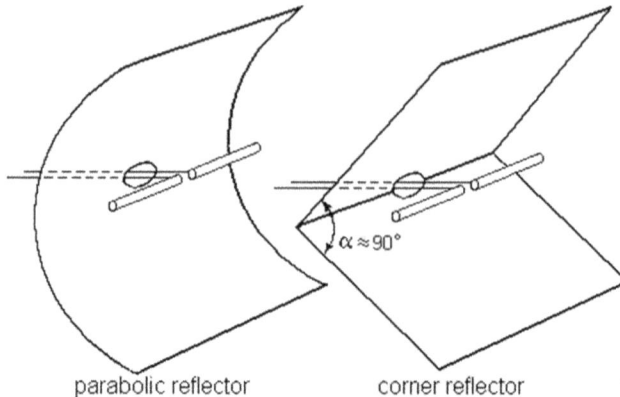

parabolic reflector   corner reflector

Figure 3.3 Reflector antenna.

3.3.4 Reflector antennas

A reflector antenna is a specific type of power transmitting or receiving antenna. Until recently, however, different geometrical forms of reflectors could not be studied or made. A reflector antenna is shown in Figure 3.3. The need for microwave communication, radio astronomy, and satellite tracking reflectors has increased substantially since the introduction of analytical and experimental techniques for reflector surfaces and the optimization of illumination over their apertures to optimum gain [16].

3.3.5 Lens antennas

Lens antennas are antennas whose direction pattern is specified by the difference in phase propagation velocities of an electromagnetic wave both in the air and in the lens material. Radar and military applications frequently employ lens antennas [13]. The beam form is defined by the shape and refractive index of the lens, the ratio of the phase velocity of the lens to the propagation in vacuum, the sizes of the three lenses are shown in Figure 3.4. An accelerating lens antenna has $n = 1$ whereas a decelerating lens has $n > 1$.

3.3.6 Yagi-Uda antennas

A Yagi-Uda antenna is a directional antenna constructed of numerous parallel metallic rods with dipole component lines. It consists of a single driven element that radiates power from the transmitter or receives power from the receiver through transmission lines, as well as a reflector in front of the driven element and one or more directional antennas.

Figure 3.5 features YAGI-UDA. In the 2.15–2.87 GHz frequency range, the antenna array performs well with a reflection coefficient of less than 10 dB. Reflectors are frequently longer than driven dipoles, although directors are usually shorter.

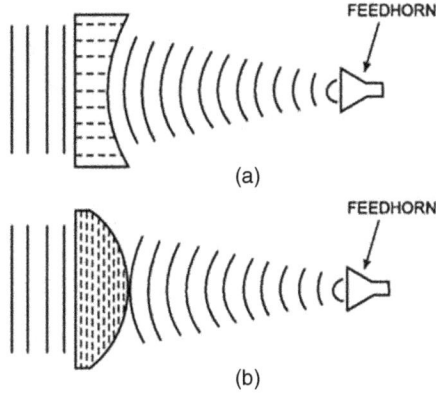

Figure 3.4 Lens antenna.

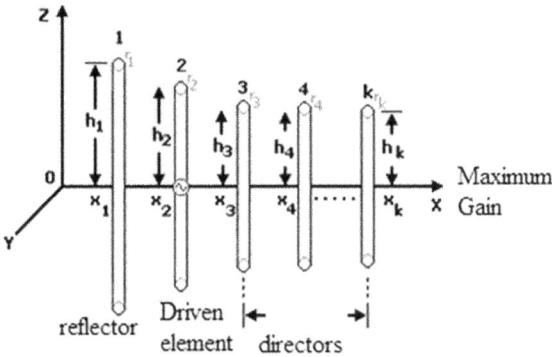

Figure 3.5 Yagi-Uda antenna array.

3.3.7 Log periodic dipole array

A Yagi-Uda antenna is a directional antenna constructed of numerous parallel metallic rods with dipole component lines. It consists of a single driven element that sends or receives power from the transmitter through transmission lines, a reflector in front of the driven element, and one or more directional antennas. Figure 3.6 shows a log periodic dipole array. Reflectors are frequently longer than driven dipoles, although directors are usually shorter [15].

The active region of an antenna wavelength dipole diminishes towards the array's beginning, where the smallest components are present. The maximum frequency is determined by the length of the shortest element.

3.3.8 Horn antennas

Horn antennas, which look like flared waveguides, are usually linked to a waveguide. It's like electromagnetic waves spreading from a corner. One of the most frequent forms of microwave antenna is the horn antenna, which

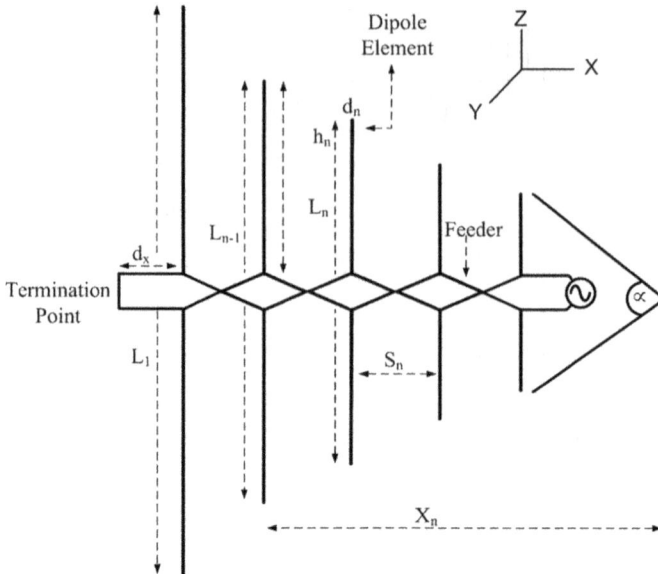

Figure 3.6 Log periodic dipole array.

was discovered and utilized for the first time in the late 1800s. It began to improve over microwave lines in the late 1930s and the early 1940s, during World War II. One of a family of artefacts detailing the radiation mechanism, growth and profitability optimization, design methodologies, and applications has emerged since then. Large radio networks, as well as satellite tracking communication dishes, use horn antennas [19]. It is a common component of phased arrays used to assess the gains of other antennas. Because of its ease of construction, ease of excitation, versatility, high gain, and overall performance, horn antennas have a wide range of applications The horn antenna is shown in Figure 3.7. The horn is a hollow metallic pipe. It might be made up of several cross-sections that taper to a huge hole.

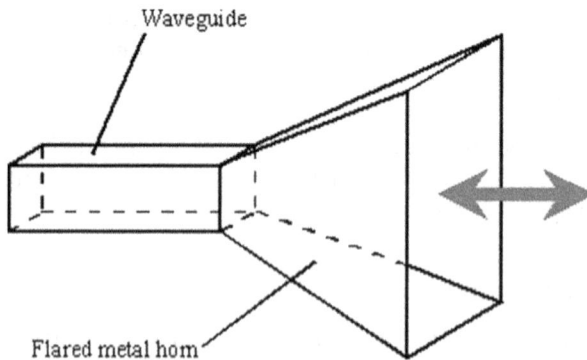

Figure 3.7 Horn antenna.

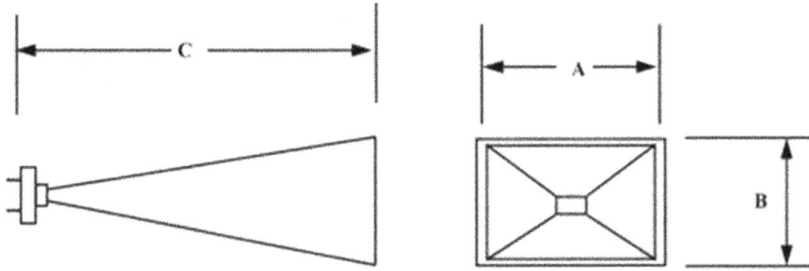

Figure 3.8 Pyramidal horn antenna dimensions (Mojave).

3.4 OBJECTIVE OF RESEARCH

The purpose of this project is to build a high-frequency gain horn antenna [20]. Anyone interested in building a typical gain horn antenna should read this part (Figure 3.8). Pyramidal horn antennas are available in a variety of shapes and sizes, and the dimensions of a given horn antenna may vary depending on its intended application, frequency range, and design specifications. The suggested study aims to develop a pyramidal horn antenna with the following characteristics.

1. Frequency range: 12 GHz to 18 GHz.
2. A increase of up to 25 decibels is achievable.
3. At the middle frequency efficiency hits 50%.
4. 4 WR62 waveguide = 0.622 0.311
5. Horn dimensions: $A = 5.60$, $B = 4.16$, $C = 12.50$

3.4.1 Microwave frequency

- Microwaves are electromagnetic beams with frequencies ranging from 300 MHz to 300 GHz.
- Microwaves are hundreds of times smaller than radio waves.
- They seldom use public transit.
- These vibrations will be totally reflected by metals.

These waves may pass through glass and particles, for example [15, 16]. Microwaves are ideal for high-bandwidth wireless communications. Microwaves are used in satellite communications, radar signals, cell phones, and navigation systems. Microwave ovens are also used in the medical profession, for food drying, and in the home kitchen. Cavity resonators or resonant lines can also be utilized to replace lumped components and tuned circuits. Even at higher frequencies, when the wavelengths of electromagnetic waves are small in contrast to the size of the structures necessary to regulate them, microwave technology and optical technology may be used. High-power microwave generators that employ specialized vacuum tubes generate microwaves.

Table 3.1 Operational frequencies

Band	Range of frequency (GHz)
L	1–2
S	2–4
C	4–8
X	8–10
High Frequency	12–18
K	18–27
Ka	27–40

3.4.2 Microwaves with band

Microwaves are separated into wavelength-based sub-bands, each with its own dataset. The various band types and their operational frequencies are shown in Table 3.1. The microwave frequency bands are as follows [15].

3.4.2.1 L-band

The frequency range of L-bands is 1 GHz to 2 GHz, with a free space wavelength of 15 cm to 30 cm. GPS navigation, GSM phones, and military applications all use these frequency ranges. They might be useful for determining the moisture content of rain forest soils.

3.4.2.2 S-band

The frequency of microwaves range from 2 GHz to 4 GHz, while the wavelengths range from 7.5 cm to 15 cm. These waves might be used in a wide range of applications.

3.4.2.3 C-band

The C-band has frequencies ranging from 4 GHz to 8 GHz and wavelengths ranging from 3.75 cm to 7.5 cm.

3.4.2.4 X-band

Microwaves have wavelengths ranging from 25 to 37.5 mm and frequency ranging from 8 to 12 GHz. Radars, broadband communications, amateur radio broadcasts, and other equipment employ these waves.

3.4.2.5 High frequency

Band frequency range of 12 to 18 GHz and a wavelength range of 16.7 to 25 mm are required. Wind speed and direction might be predicted using microwave pulse energy, which is employed in satellite communications.

3.4.2.6 K-band and Ka-band

K-band waves have a frequency range of 18 to 26.5 GHz and wave heights that range from 11.3 and 16.7 mm. The Ka-band has a frequency range of 26.5 GHz to 40 GHz. Satellite communications, radars, and astronomical research all make use of these waves. Radars in this frequency band feature a high data rate, a limited range, and exceptional resolution.

3.5 HORN ANTENNA

Horn antennas are commonly used in UHF and higher microwave bands. Horn antennas are recognised by their directed emission pattern and high antenna gain, which normally ranges from 10 to 20 dB but can reach 25 dB on rare occasions. Horn antennas have a wide impedance bandwidth, which means their input impedance varies slowly across a wide frequency range. This results in a low voltage standing wave ratio (VSWR) over the entire bandwidth.

Practical horn antennas have a 10:1 bandwidth and operate in the 20:1 range. Horn antennas frequently get additional gain as the operating frequency is raised. Because of the shorter wavelengths, higher frequencies have "electrically larger" horn antennas. This is because horn aperture data are frequently reported in wavelengths. Because of its constant physical size, the horn antenna can withstand bigger apertures at higher frequencies. Antennas are simple to construct and offer a high level of intuitiveness. Acoustic horn antennas may also transmit sound waves. Horn antennas are widely used for measurements or to feed a dish antenna as a "standard gain" antenna [17].

3.5.1 Waveguides

A waveguide is a structure that confines energy transmission to one direction and so directs waves with minimum energy loss, such as an acoustic waveguide, an optical waveguide, radio waves and radio-frequency waveguide. Rather than transmitting a signal, a waveguide serves as a unit for channelling and controlling the flow of electromagnetic radiation. Because it is a single-conductor system, its electrical distribution characteristics differ from those of two-conductor transmission lines. Waveguides are generally made of metal hollow tubes. They can withstand massive power loads while also directing electricity precisely where it is required and acting as a high-pass filter. At microwave frequencies, waveguides are commonly employed. Wideband waveguides may carry or transmit power and data signals. A rectangular hollow metal waveguide is shown in Figure 3.9. The mode of an electromagnetic wave is the arrangement of its electric and magnetic field components [10, 8].

Figure 3.9 Image of a rectangular waveguide.

3.5.2 Types of wave-guide

3.5.2.1 Circular waveguides

In radars, circular waveguides are used in conjunction with rotating antennas to twist waves as they pass through them.

3.5.2.2 Elliptical waveguides

Elliptical waveguides are a kind of waveguide commonly used in flexible waveguides. These waveguides will be required if a portion of the waveguide is capable of moving, such as bending, stretching, or twisting.

3.5.2.3 Ridged waveguides

Ridged waveguides feature conducting ridges that extend from the top, bottom, or both walls into the core. They can be used to provide impedance matching and to increase bandwidth [5, 6]. Other waveguides are not included on this page since they have not been properly researched. Figure 3.10 depicts many waveguide types the next section goes into great depth on rectangular waveguides.

3.5.2.4 Rectangular waveguide

Rectangular cross-section waveguides are ideal for studying electrodynamics forces in three dimensions. A waveguide is a structure that confines electromagnetic impulses, lowering dispersion, losses, and signal transmission from one site to another. A hollow conducting tube is commonly used to construct a basic waveguide. When the conducting tube has a rectangular cross-section, a rectangular waveguide is generated. The next section will cover a range of issues in rectangular waveguide theory [7, 16]. A rectangular waveguide is the most common form of waveguide. They are hollow metallic structures with rectangular sections. A rectangular waveguide is

Types of Waveguides

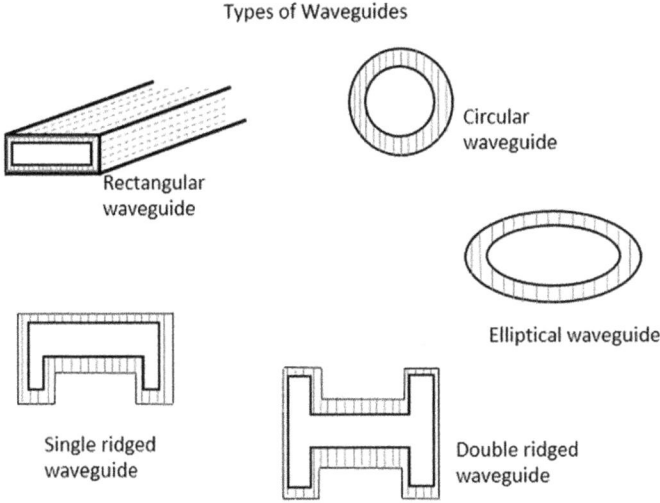

Figure 3.10 Types of waveguide.

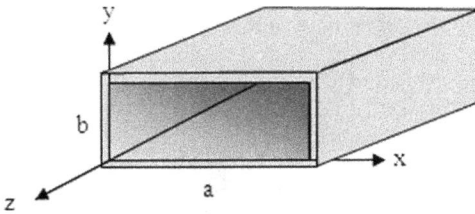

Figure 3.11 Rectangular waveguide.

frequently constructed with a length larger than b, where b is the rectangle's width. [15]. As a result, for radiation at 3 GHz, a 5 cm diameter guide is commonly utilized (Figures 3.11 and 3.12). Aspect ratio-dependent normal-ized cutoff frequencies for lowest rectangular waveguide modes.

In a rectangular waveguide, both the TE and TM modes can propagate. The subscripts m and n are often used to signify TE and TM modes. The symbols TMmn and TEmn are used to represent TMmn [22, 16]. The full numbers m and n represent the strength of TE and TM modes between each pair of barriers. The m is measured along the waveguide's x-axis (dimension a), which is the bigger wall dimension; the n is measured along the y-axis.

Cut-off frequency: The cut-off frequency of an EM waveguide is the low-est frequency at which a mode will propagate in it.

$$f_c = \frac{c}{2}\left[\left(\frac{m}{a}\right)^2 + \left(\frac{n}{b}\right)^2\right]^{\frac{1}{2}}$$

(3.11)

TE$_{10}$

TM$_{11}$

TE$_{11}$

----- E field
----- H field

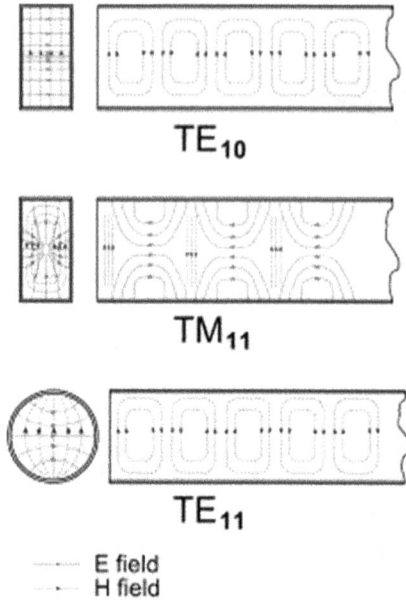

Figure 3.12 Rectangular waveguide modes.

$$\lambda_c = \frac{2ab}{\sqrt{m^2 b^2 + n^2 a^2}} \tag{3.12}$$

Dominant mode: The lowest dominant mode in a waveguide has a cut-off frequency of TE10

$$\lambda_{c(10)} = 2a \tag{3.13}$$

3.6 HORN ANTENNA DESIGN FORMULA

3.6.1 Horn antenna design

Losses are reduced and beam focusing is improved when the energy in the beam is gradually converted to radiation. A horn antenna may be thought of as a wave path that has been fanned out to improve directivity and reduce diffraction [10, 17]. Higher frequencies are served by horn antennas. Horn antennas have a concentrated emission pattern and a high gain Figure 3.13 shows the design of a pyramidal horn antenna flaring a rectangular horn in both the E and H planes results in a pyramidal horn antenna.

Equations 3.14 and 3.15 illustrates how the E and H field across the aperture of a horn is approximated by

$$E'_y\left(x^-, y^-\right) = E_0 \cos\cos\left(\frac{\pi}{a_1} x^-\right) \exp\left[\frac{-jk}{2}\left\{\frac{x^-}{\rho_2} + \frac{-}{\rho_1}\right\}\right] \tag{3.14}$$

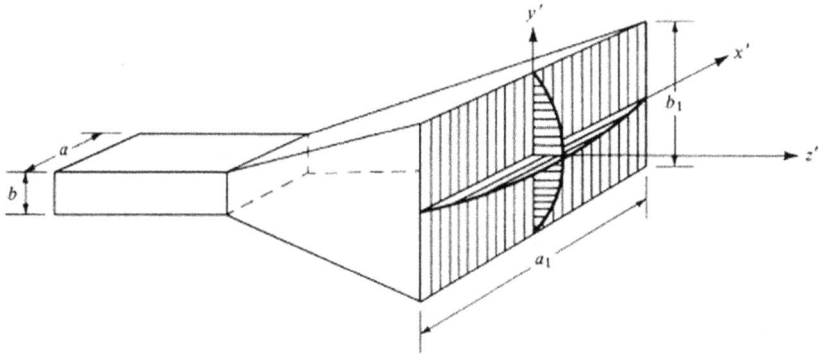

Figure 3.13 Pyramidal horn antenna design, (a) E-plane view, (b) H-plane view.

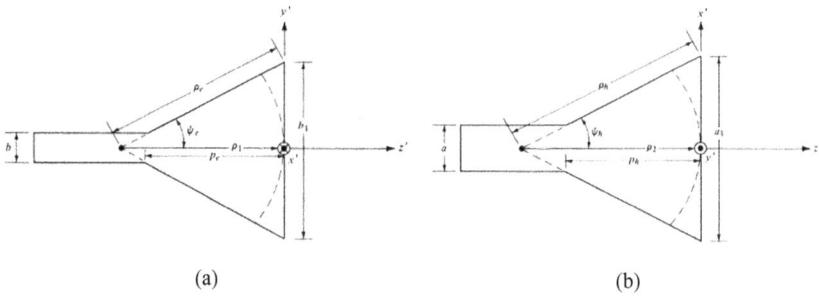

(a)   (b)

Figure 3.14 E-plane & H-plane view.

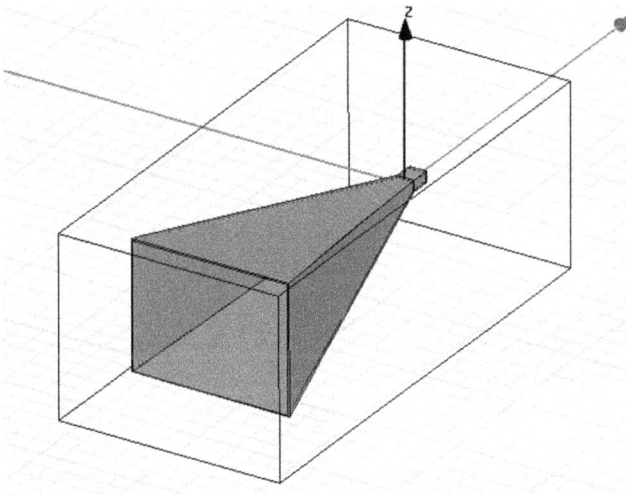

Figure 3.15 Designed pyramidal horn antenna.

$$H'_x\left(x^-,y^-\right)=\frac{-E_0}{\eta}\cos\cos\left(\frac{\pi}{a_1}x^-\right)\exp\left[\frac{-jk}{2}\left\{\frac{x^-}{\rho_2}+\frac{-}{\rho_1}\right\}\right]\qquad(3.15)$$

The radiated wave is represented by the exponential portion. The length of the horn's aperture area defines it. Figure 3.14 shows the E-plane and H-plane views.

The actual realization of a horn antenna is determined as follows:

$$\rho_e = \rho_h \qquad (3.16)$$

Gain is generally one of the beginning areas for horn antenna design:

$$G_0 = \frac{1}{2}\left(\frac{4\pi}{\lambda^2}a_1b_1\right) \qquad (3.17)$$

where
a_1 is the length of the horn flared mouth and
b_1 is the breadth of the horn flared mouth.

$$a_1 \approx \sqrt{3\lambda p_2} \approx \sqrt{3\lambda p_h}$$

$$\left(p_2 \approx p\right) \qquad (3.18)$$

$$b_1 \approx \sqrt{2\lambda \rho_1} \approx \sqrt{2\lambda p_e}$$

$$\left(\rho_1 \approx p_e\right) \qquad (3.19)$$

A common profit the high-frequency horn antenna is built with the waveguide. The specified gain is 25 dB and the frequency range is 12.4 to 18 GHz.

3.6.2 Horn dimension

The size of a horn antenna can vary substantially, depending on its design and intended function. Horn antennas are commonly used in microwave and radio frequency communication systems, and their size is often determined by the operating frequency and desired emission pattern [13, 20]. The throat width is the width of the opening at the horn's narrow end.

It is commonly stated as a wavelength or as a percentage of the operating wavelength. The width of the mouth is defined as the width of the aperture at the wide end of the horn, that is, the throat width. Figure 3.16 depicts the designed pyramidal horn antenna.

The waveguide length and breadth of a standard horn antenna Wavelength and waveguide dimension calculation

E-plane view

H-plane view

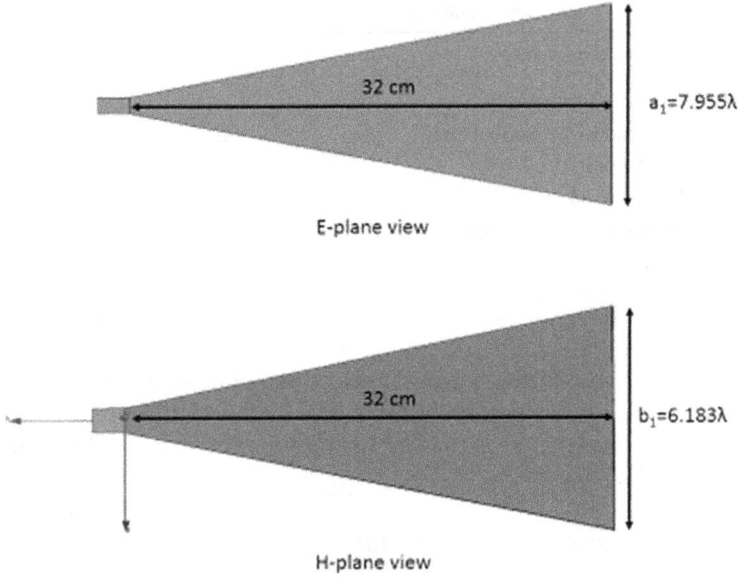

Figure 3.16 **E**-planes and **H**-plane view of horn.

3.6.3 Design procedure of horn antenna

The optimization design parameter

$$\chi_1 = \frac{G_0}{2\pi\sqrt{2\pi}} \tag{3.20}$$

$$\chi_1 = \frac{316.23}{2\pi\sqrt{2\pi}} = 20.078 \tag{3.21}$$

To satisfy the equation below, the value of X_1 must be optimized.

$$\frac{\left(\sqrt{2\chi} - \dfrac{b}{\chi}\right)}{2\chi - 1} = \left(\frac{G_0}{2\pi}\sqrt{\frac{3}{2\pi}}\frac{1}{\sqrt{\chi}} - \frac{a}{\lambda}\right)^2\left(\frac{G_0^2}{6\pi^3}\frac{1}{\chi} - 1\right) \tag{3.22}$$

3.7 SIMULATION RESULTS

The size of a horn antenna can vary substantially depending on its design and intended function. Horn antennas are commonly used in microwave and radio frequency communication systems, and their size is often determined by the operating frequency and desired emission pattern. HFSS 11 was used to simulate a horn antenna. The introduction of the horn antenna was necessary due to the failure of auto mesh sizing to function [22]. It has diameters ranging from 0.7 to 32 cm. Table 3.2 shows the range of frequency and

Table 3.2 Range of frequency

Antenna parameter	Values
Operating Frequency	12 GHz to 18 GHZ
Bandwidth For Return Loss	4 GHz
Bandwidth For VSWR	4 GHz
Maximum Gain	25 dB
Minimum Return Loess	−40 dB
Radiation Pattern	Highly Directive
Minimum VSWR	1.02 at 15.8 GHz

how the simulation duration grows as the mesh dimension is increased. The frequency is maintained above 10 GHz. The mesh in HFSS may be adjusted based on the length of the area. The smallest mesh size is 0.4 cm, while the largest mesh size is 7 cm.

3.7.1 Reflection coefficient and VSWR

The reflection coefficient in physics and electrical engineering measures how much of a wave is reflected by an impedance discontinuity in the transmission medium. It equals the amplitude ratio of the incident to reflected waves, which are both represented as phases [22]. Return loss is one of the scattering parameters (S11). The quantity of power returned from a transmission line is indicated by the return loss value. The load in this case is an antenna with the highest negative value for the predicted frequency of operation. Figure 3.17 displays the proposed horn antenna's return loss versus frequency curve.

The Voltage Standing Wave Ratio (VSWR) is the maximum voltage to lowest voltage ratio of a load-supplied gearbox line. An antenna's VSWR reveals its ability to match impedance. Our antenna has a VSWR of less than 1.15 over the operational range stated, which is less than one and less than two. Figure 3.18 depicts the VSWR of the proposed antenna.

An antenna's bandwidth may be estimated using both the return loss and VSWR charts. The VSWR bandwidth is defined as the area of the plot that lies below a certain VSWR value [15, 21]. Methods for enhancing return loss and VSWR bandwidth are depicted in Figures 3.17 and 3.18. The frequency band is between 12 and 18 GHz and the frequency range of the antenna is 12 to 18 GHz. This frequency band is mostly employed in satellite communications. The antenna's bandwidth percentage is 26.67%.

$$\text{BW}\% = \frac{f_2 - f_1}{f_c} \times 100 \tag{3.23}$$

$$\text{BW}\% = \frac{18 - 12}{15} \times 100 = 26.67\%$$

A −20 dB reference is used to determine f_1 and f_2.

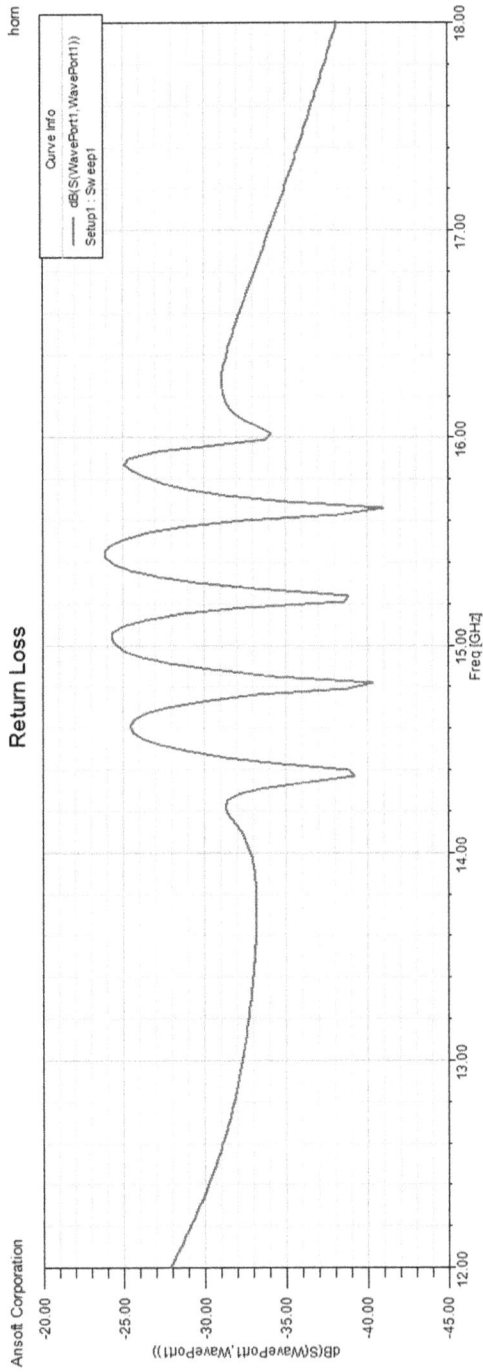

Figure 3.17 Return loss vs frequency plot.

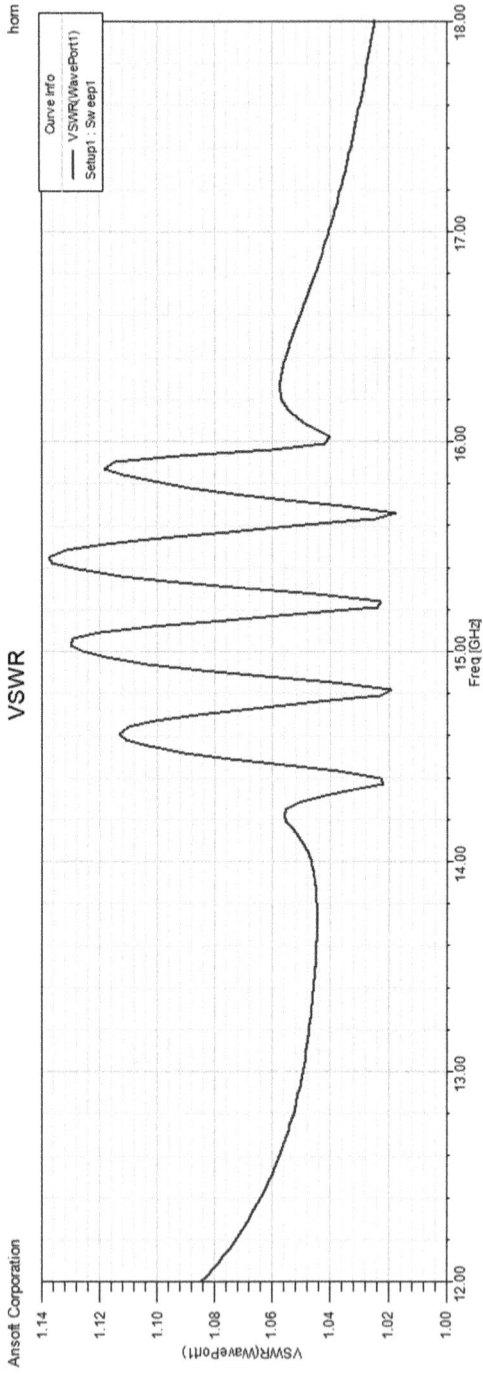

Figure 3.18 VSWR vs frequency plot.

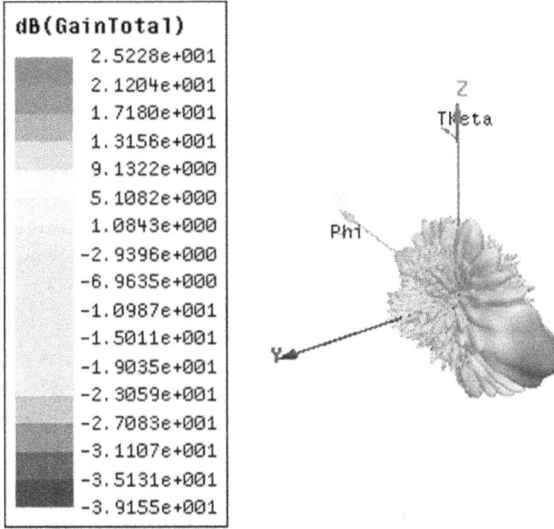

Figure 3.19 3D radiation pattern.

3.7.2 Radiation pattern

A three-dimensional radiation pattern is depicted in Figure 3.19. The radiation pattern of an antenna demonstrates how the antenna emits radiation. It indicates the location of the antenna. The radiation pattern is vertically polarized and stimulated in the TE10 mode. Figures 3.19 and 3.20 show 3D Radiation Patterns and 2D Radiation Patterns, respectively. There are several side lobes in addition to the main beam.

3.7.3 Gain and directivity

Figures 3.21 and 3.22 show the antenna's gain vs frequency plot and directivity vs frequency plot. These show the radiation pattern of the H and E planes. The width of the H-plane beam is 6.88 degrees, while the width of the E-plane beam is 9.98 degrees.

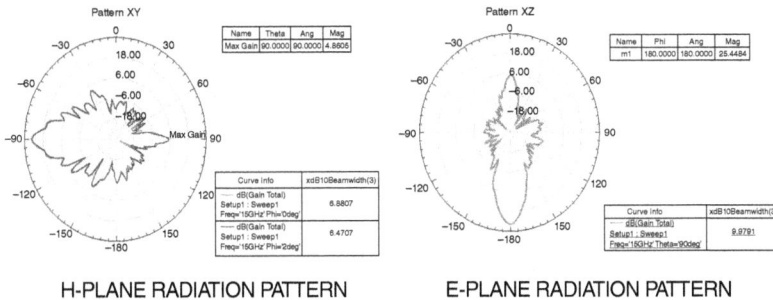

H-PLANE RADIATION PATTERN   E-PLANE RADIATION PATTERN

Figure 3.20 2D radiation pattern.

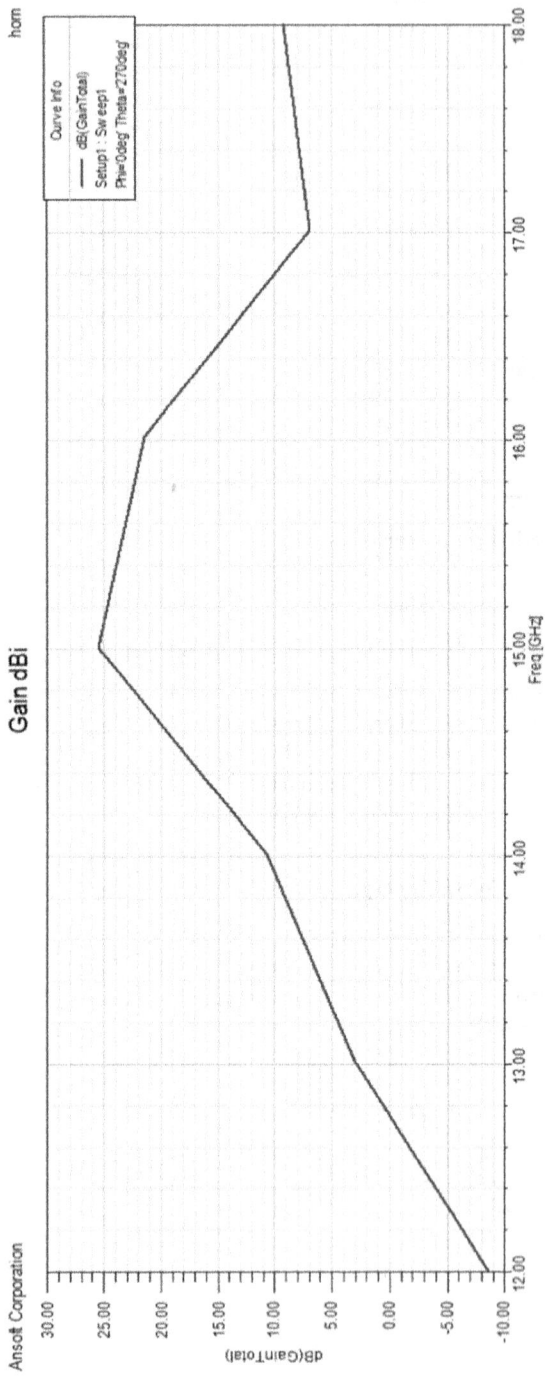

Figure 3.21 Gain vs frequency plot.

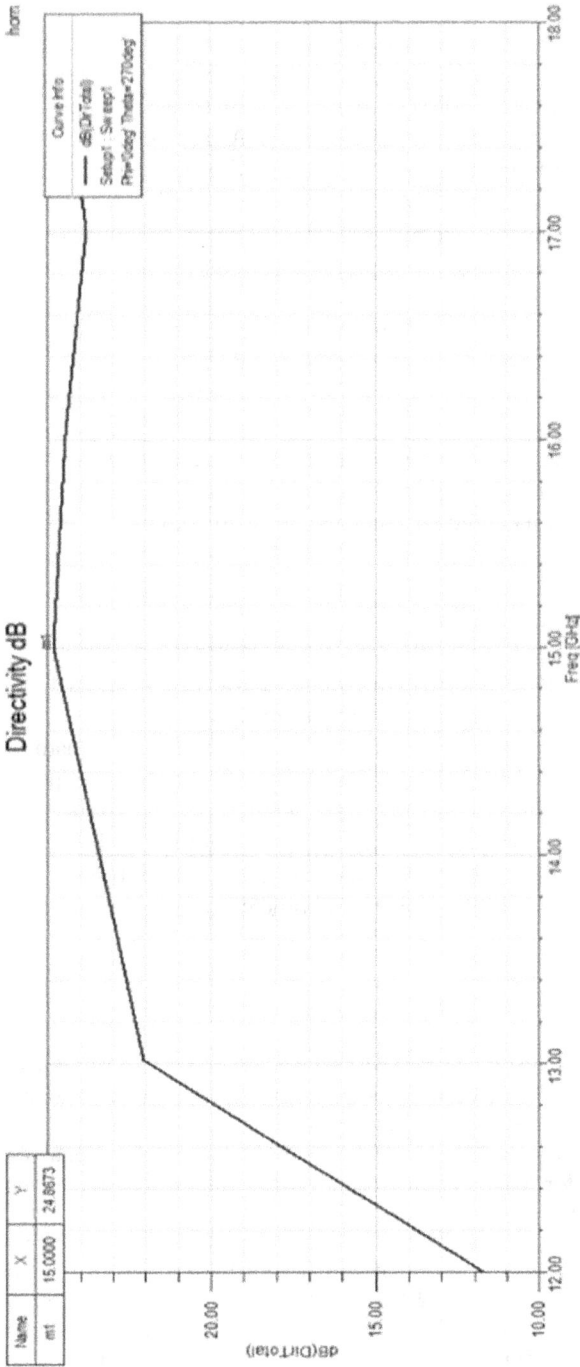

Figure 3.22 Directivity vs frequency plot.

3.8 CONCLUSION

A horn antenna is a type of aperture antenna designed particularly for microwave frequencies. The antenna tip is either widened or fashioned like a horn. Because of the enhanced directivity of this structure, the resultant signal may be easily transmitted across long distances. A directional antenna is usually accompanied with a waveguide known as a horn antenna, which works by using the microwave spectrum. Because coaxial cable feeding has a higher reflection coefficient, future research should focus on optimal feeder design. The usual effectiveness of an antenna is roughly 50%. The antenna's bandwidth of 30% is outstanding. If constructed, the suggested antenna can be used as a standard gain antenna. This horn can be used as a feed for a parabolic dish or as microwave testing equipment. Because coaxial cable feeding has a greater reflection coefficient, future study will focus on optimal feeder design. The suggested antenna can be used as a standard gain antenna. In a microwave testing facility, this horn can be used as test equipment or as the feed for a parabolic dish. Future study will focus on the ideal feeder design when the reflection coefficient increases with coaxial cable feeding.

3.9 FUTURE ASPECTS

The horn antenna is an aperture antenna made in particular for microwave frequencies. In such antennas, the tip is either broader or horn-shaped. The greater directivity of this structure allows for the easy transmission of the generated signal across long distances. To enhance performance and reduce phase error, the horn antenna's flare length may be changed in order to reduce the phase inaccuracy to between 450 and 900. Investigations may also focus on different shapes, such as dual ridged horns, stepped horn antennas, corrugated horn antennas, conical horn antennas, etc.

REFERENCES

1. Naser Ojaroudi Parchin, and Heba G. Mohamed "An Efficient Antenna System with Improved Radiation for Multi-Standard/Multi-Mode 5G Cellular Communications," *Sci. Rep.*, vol. 13, no. 1, pp. 1–11, 2023.
2. Jinhyeok Park, and Songcheol Hong, "Wideband Bidirectional Variable Gain Amplifier for 5G Communication," *IEEE Microw Wirel Compon Lett.*, vol. 33, no. 6, pp. 691–694, 2023.
3. Youngmin Kim, and Sangmin Yoo, "High-Efficiency 28-/39-GHz Hybrid Transceiver Utilizing Si CMOS and GaAs HEMT for 5G NR Millimeter-Wave Mobile Applications," *IEEE Solid-State Circuits Lett.*, vol. 6, pp. 1–4, 2023.
4. Xiaosong Liu, and Zehong Yan "Dual-Band Orthogonally-Polarized Dual-Beam Reflect-Transmit-Array With a Linearly Polarized Feeder," *IEEE Trans. Antennas Propag.*, vol. 70, nó. 9, pp. 8596–8601, 2022.

5. Mohammed Amine Zafrane, and Bachir Abes, "Novel Design and Optimization of S Band Patch Antenna for Space Application by Using a Gravitational Search Algorithm," *Int. J. Interact. Des. Manuf.*, vol. 17, no. 3, pp. 1131–1148, 2022.

6. Vinita Mathur, and Suman Sharma, "Bridged Concentric Circular Microstrip Patch Antenna for C, X and High Frequency Band Applications," *Mater. Today: Proc.*, vol. 46, no. 7, pp. 5742–5747, 2022.

7. Rahul High, and Amit Birwal, "Design and Analysis of Compact Triple band Microstrip Patch Antenna," *2022 2nd International Conference on Emerging Frontiers in Electrical and Electronic Technologies (ICEFEET)*, pp. 1–6, 2022.

8. M. Praneetha, K. Naveen, "Tongs Shaped Dual Band MIMO Antenna for WLAN and WiMAX Applications," *2022 IEEE 10th Region 10 Humanitarian Technology Conference (R10-HTC)*, pp. 403–407, 2022.

9. Libin Sun, Yue Li, Zhijun Zhang, "Wideband Dual Polarized Endfire Antenna Based on Compact Open Ended Cavity for 5G Mm Wave Mobile Phones" *IEEE Trans. Antennas Propag.*, vol 70, no. 3, pp. 1632–1642, Mar. 2022.

10. Benoit Brizard "Dual High Frequency -Band Dielectric Resonator Antenna Sub-Array Fed by a Substrate Integrated Coaxial Line," *2022 IEEE USNC-URSI Radio Science Meeting*, pp. 48–49, 2022.

11. Inzamam Ahmad, and Ernesto Limiti, "Design and Analysis of a Photonic Crystal Based Planar Antenna for THz Applications," *Electronics*, vol. 10, no. 16, 2021. doi:10.3390/electronics10161941

12. Zhechen Zhang, and Shiwen Yang, "In-Band Scattering Control of Ultra-Wideband Tightly Coupled Dipole Arrays Based on Polarization-Selective Metamaterial Absorber," *IEEE Trans. Antennas Propag.*, vol. 68, no. 12, pp. 7927–7936, 2020.

13. Adelaida Heiman "Design of a Conventional Horn Antenna for High Frequency Band," In *2020 International Workshop on Antenna Technology (iWAT) Date of Conference*, 25–28 February 2020.

14. Biao Hao, and Tong Li, "A Coding Metasurface Antenna Array with Low Radar Cross Section," *Acta PhysicaSinica*, vol. 69, no. 24, pp. 244101, 2020.

15. Neng-Wu, and Guang Fu, "Frequency-Ratio Reduction of a Low Profile Dual Circular Polarized Patch Antenna Under Triple – Resonance," *IEEE Antennas and Wireless Propagation Letters*, vol. 19, no. 10, pp. 1689–1693, Oct. 2020.

16. Muhammad Saleem, and Xiao-Lai Li, "Low Scattering Microstrip Antenna Based on Broadband Artificial Magnetic Conductor Structure," *Materials*, vol. 13, no. 3, pp. 750, 2020.

17. Syah Alam, and Aida Safitri, "Design of Truncated Microstrip Antenna with Array 4×2 for Microwave Radio Communication," *2019 IEEE Conference on Antenna Measurements & Applications (CAMA)*, pp. 1–4, 2019.

18. Hossein Vahimian, and Nasrin Amiri, "Multi-band High Impedance Surfaces Using Progressively Magnified EBG Structures," *2019 Sixth Iranian Conference on Radar and Surveillance Systems*, pp. 1–5, 2019.

19. Sai Radavaram, and Maria Pour, "Wideband Radiation Reconfigurable Microstrip Patch Antenna Loaded With Two Inverted U Slot," *IEEE Trans. Antennas Propag.*, vol. 67, no. 3, pp. 1501–1508, Mar. 2019.

20. Joysmita Chatterjee, and Vivek Dixit, "Broadband Circular Polarized H-Shaped Patch Antenna Using Reactive Impedance Surface," *IEEE Antenna and Wireless Propagation Letters*, vol. 17, no. 4 pp. 625–628, Apr. 2018

21. Hang Wong, and Xia Gao, "Bandwidth Enhancement of a Monopolar Patch Antenna with V-Shaped Slot for Car to Car and WLAN Communication," *IEEE Trans. Veh. Technol.*, vol. 2016, pp. 01–07, 2016.
22. M. Li et al. "Eight-Port Dual-Polarized MIMO Antenna for 5G Smartphone Applications," In *2016 IEEE 5th Asia-Pacific Conference on Antennas and Propagation (APCAP)*, pp. 195–196, 2016.

Chapter 4

Antenna design for IoT and 5G applications

Advancements, challenges, and future perspectives

Pritesh Kumar Jain and Sandeep Kumar Jain

Shri Vaishnav Vidyapeeth Vishwavidyalaya, Indore, India

4.1 INTRODUCTION

The rapid proliferation of wireless communication technologies has given rise to the Internet of Things (IoT) and the imminent deployment of fifth-generation (5G) networks. These advancements have revolutionized the way devices and systems interact, enabling seamless connectivity and ubiquitous access to information. However, to fully realize the potential of IoT and 5G, efficient and versatile antenna designs are essential. Antennas play a critical role in enabling reliable wireless communication, and their design and performance directly impact the overall system capabilities.

The design of antennas for IoT and 5G applications presents unique challenges compared to traditional wireless systems [1]. IoT devices are characterized by their small form factors, low power consumption, and diverse operating environments. These devices encompass a wide range of applications, from smart homes and wearable to industrial sensors and autonomous vehicles. As such, IoT antennas need to be compact, energy-efficient, and capable of operating across multiple frequency bands to accommodate various connectivity requirements.

On the other hand, 5G networks demand antennas that can support the increased data rates, improved spectral efficiency, and enhanced capacity needed to deliver the promised transformative capabilities [2]. These networks utilize higher frequency bands, including millimeter waves, to achieve the desired performance levels. Consequently, 5G antennas must address challenges related to signal propagation, beam-forming, and multi-user interference to enable reliable and high-speed wireless connectivity.

In the context of IoT and 5G applications, antenna design parameters such as radiation pattern, gain, efficiency, polarization, and impedance matching become crucial. Antennas must exhibit desirable characteristics, such as omnidirectional or directional radiation patterns, high gain to enhance signal strength, high efficiency to minimize power consumption, polarization diversity to accommodate varying device orientations, and impedance matching to ensure maximum power transfer.

DOI: 10.1201/9781003422440-4

To meet these requirements, recent advancements in antenna design have focused on various innovative techniques [3, 4]. Small-sized and multi-band antennas have been developed to cater to the compact form factor and multi-frequency operation needs of IoT devices. Phased array antennas and beam-forming techniques have gained prominence in 5G systems, enabling the formation of highly directional beams for improved link quality and spectral efficiency. Additionally, reconfigurable antennas offer flexibility by dynamically adjusting their characteristics based on the communication requirements.

However, despite these advancements, antenna design for IoT and 5G applications presents several challenges. Miniaturization is a crucial consideration, as IoT devices often have limited space available for antenna integration. Bandwidth efficiency is another concern, particularly in 5G networks, where the demand for high data rates requires antennas to operate over wide frequency ranges. Electromagnetic interference and coexistence issues arise due to the proliferation of wireless systems, necessitating careful design and shielding techniques. Furthermore, regulatory constraints must be considered to ensure compliance with spectrum allocation regulations and safety limits.

Looking ahead, the future of antenna design for IoT and 5G applications holds significant promise. Millimeter-wave communications, massive multiple-input multiple-output (MIMO), and intelligent antennas are emerging technologies that aim to further enhance wireless system performance. Research efforts are underway to explore these areas and develop novel antenna designs that push the boundaries of efficiency, capacity, and reliability. Moreover, interdisciplinary collaborations between antenna designers, wireless communication experts, and standardization bodies are vital for driving innovation and ensuring seamless integration of IoT and 5G technologies.

In summary, antenna design plays a crucial role in enabling efficient wireless communication for IoT and 5G applications. As the demand for connectivity and data exchange continues to grow, antennas must meet the unique requirements of these applications, including small form factors, multi-frequency operation, and high data rates. Advancements in antenna design techniques offer promising solutions, but challenges related to miniaturization, bandwidth efficiency, electromagnetic interference, and regulatory compliance must be addressed. By exploring emerging technologies and fostering collaborations, researchers and engineers can drive further.

4.2 TYPES OF ANTENNA USED IN IOT AND 5G

Various types of antennas are utilized in IoT and 5G applications to cater to specific requirements and optimize wireless communication performance. Here are some commonly used antennas in these domains [5–7]:

4.2.1 Monopole antennas

Monopole antennas are widely used in IoT devices and 5G systems due to their compact size and omnidirectional radiation patterns. They consist of a vertical conducting element mounted over a ground plane, making them suitable for applications where space is limited, such as wearables, smart home devices, and small IoT sensors.

4.2.2 Patch antennas

Patch antennas are planar antennas that consist of a metal patch mounted over a ground plane. They offer advantages such as low profile, ease of integration, and good radiation characteristics. Patch antennas are commonly employed in IoT devices, such as smart tags, tracking devices, and wireless sensor nodes, where they can provide directional or wide-angle coverage.

4.2.3 Dipole antennas

Dipole antennas consist of a conductive element that is split into two halves and fed at the center. They are known for their omnidirectional radiation pattern and are widely used in IoT and 5G applications. Dipole antennas are employed in applications where a balanced radiation pattern is required, such as wireless routers, access points, and indoor IoT devices.

4.2.4 Helical antennas

Helical antennas are compact, high-gain antennas that operate over a wide frequency range. They consist of a wire wound in the form of a helix and are often used in IoT and 5G applications that require circular polarization, such as satellite communication, RFID systems, and wireless positioning systems.

4.2.5 Yagi-Uda antennas

Yagi-Uda antennas are highly directional antennas commonly used in long-range IoT applications and point-to-point communications. They consist of multiple dipole elements, with one being the driven element and the others acting as directors and reflectors. Yagi-Uda antennas offer high gain and improved signal reception in specific directions.

4.2.6 Planar inverted-F antennas (PIFA)

PIFA antennas are compact, planar antennas that are commonly integrated into smartphones, tablets, and IoT devices. They offer wideband operation, low profile, and good radiation characteristics. PIFA antennas are suitable for applications that require compact size and good performance in wireless communication.

4.2.7 Phased array antennas

Phased array antennas are an advanced type of antenna used in 5G systems. They consist of multiple antenna elements that can be electronically controlled to steer the beam direction without physically moving the antenna. Phased arrays offer beam-forming capabilities, allowing for the formation of directional beams and improved signal quality.

4.2.8 Multiple-input multiple-output (MIMO) antennas

MIMO antennas involve the use of multiple antennas at both the transmitter and receiver ends to enhance the capacity and reliability of wireless communication. MIMO technology is widely employed in 5G systems, enabling higher data rates, increased spectral efficiency, and improved signal robustness.

These are just a few examples of the antenna types used in IoT and 5G applications. The choice of antenna depends on factors such as application requirements, frequency bands, size constraints, radiation pattern needs, and desired performance parameters. As technology continues to evolve, new antenna designs and configurations are being developed to meet the specific demands of IoT and 5G networks.

4.3 PERFORMANCE PARAMETERS

Performing a comparative analysis of performance parameters between IoT and 5G applications involves evaluating key metrics that determine the efficiency, reliability, and capabilities of the respective systems. Here are some performance parameters commonly considered [8, 9]:

4.3.1 Data rate

Data rate refers to the speed at which data can be transmitted or received in a given system. In general, 5G networks offer significantly higher data rates compared to IoT applications. 5G promises peak data rates in the multigigabit per second range, enabling seamless streaming, high-quality video calls, and real-time data-intensive applications. IoT applications typically have lower data rate requirements, as they often involve transmitting small packets of data intermittently.

4.3.2 Latency

Latency is the delay between a data transmission request and the receipt of the corresponding response. In both IoT and 5G, low latency is crucial for real-time applications and interactive services. 5G networks aim to achieve ultra-low latency in the millisecond range, enabling responsive applications such as autonomous vehicles, remote surgery, and immersive virtual reality. IoT applications typically have more relaxed latency requirements, but low-latency communication is still important for time-critical IoT deployments like industrial automation and healthcare monitoring.

4.3.3 Coverage

Coverage refers to the geographical area within which a wireless network can provide reliable connectivity. 5G networks aim to offer broader coverage compared to traditional cellular networks, utilizing various frequency bands, including low-frequency sub-6 GHz bands and high-frequency millimeter waves. IoT networks, on the other hand, can operate over a range of coverage areas, from localized deployments (e.g., smart homes, industrial sites) to wide-scale deployments (e.g., smart cities). IoT coverage can be achieved through a combination of short-range wireless technologies like Bluetooth, Zigbee, and Wi-Fi, as well as long-range technologies like cellular IoT (LTE-M, NB-IoT).

4.3.4 Device density

Device density refers to the number of devices that can be supported within a given area or cell. IoT applications often require high device density, as they involve large-scale deployments of interconnected devices. IoT networks are designed to handle massive numbers of devices, from thousands to millions, operating in close proximity. 5G networks are also built to support high device densities, particularly in scenarios like smart cities, where numerous connected devices, sensors, and infrastructure components need to coexist and communicate seamlessly.

4.3.5 Power consumption

Power consumption is a critical factor in IoT applications, where devices often operate on battery power for extended periods. IoT systems are designed to be energy-efficient, optimizing power consumption to prolong battery life. In contrast, 5G networks, being infrastructure-based, have more flexibility in terms of power supply. However, power efficiency is still an important consideration for 5G base stations and network equipment to minimize operational costs and environmental impact.

4.3.6 Scalability

Scalability refers to the ability of a system to handle increasing demand and accommodate growth. IoT networks are designed to scale horizontally, allowing for the seamless addition of new devices and expansion of the network infrastructure as the number of connected devices increases. 5G networks are also designed with scalability in mind, supporting the anticipated growth in data traffic and the increasing number of connected devices.

It is important to note that the specific performance parameters and their significance may vary depending on the use case and application requirements within IoT and 5G domains. The comparative analysis should consider these parameters while aligning them with the specific needs and goals of the applications being evaluated.

4.4 ANTENNA DESIGN PARAMETER IN CONTEXT OF IOT AND 5G

In the context of IoT and 5G, several antenna design parameters [10, 11] are crucial to ensure efficient and reliable wireless communication. These parameters help optimize the antenna's performance and its compatibility with the specific requirements of IoT and 5G applications. Here are some key antenna design parameters.

4.4.1 Radiation pattern

The radiation pattern of an antenna describes the distribution of electromagnetic energy in space. For IoT and 5G applications, antennas with specific radiation patterns are often desired, such as omnidirectional (covering all directions equally), directional (concentrating energy in a specific direction), or sectorial (covering a specific sector). The radiation pattern determines the coverage area, signal strength, and interference characteristics.

4.4.2 Gain

Gain refers to the measure of an antenna's ability to focus electromagnetic energy in a specific direction. Higher gain antennas are desirable in situations where long-range communication or focused coverage is required. IoT and 5G applications may require antennas with specific gain characteristics to optimize signal transmission and reception.

4.4.3 Bandwidth

Bandwidth is the range of frequencies over which an antenna can efficiently operate. In IoT and 5G applications, wide bandwidth antennas are desirable to support the multiple frequency bands used in these systems. The ability to cover various frequency bands ensures compatibility with different communication standards and maximizes the antenna's versatility.

4.4.4 Efficiency

Efficiency measures the ability of an antenna to convert electrical power into radiated electromagnetic energy. High efficiency antennas are essential for IoT and 5G applications as they help minimize power consumption and maximize the utilization of available power resources. Efficient antennas also contribute to better signal quality and overall system performance.

4.4.5 Polarization

Polarization refers to the orientation of the electric field vector in an electromagnetic wave radiated by an antenna. Matching the polarization between the transmitting and receiving antennas is crucial for optimal signal reception. Antennas designed for IoT and 5G applications need to consider the specific polarization requirements, such as linear, circular, or dualpolarization, to ensure reliable and efficient communication.

4.4.6 Impedance matching

Impedance matching is the process of matching the impedance of the antenna with that of the transmission line or the system to which it is connected. Proper impedance matching minimizes signal reflections and ensures efficient power transfer between the antenna and the system, maximizing the signal strength and minimizing losses.

4.4.7 Size and form factor

IoT devices often have stringent size constraints, requiring antennas to be compact and integrated into small form factors. Similarly, 5G systems may require antenna arrays or compact designs to accommodate multiple antennas in a limited space. Antenna size and form factor considerations are crucial to enable the seamless integration of antennas into IoT devices and 5G infrastructure.

These antenna design parameters are critical to optimize the performance, coverage, and compatibility of antennas in IoT and 5G applications. Careful consideration of these parameters ensures efficient wireless communication, enhances system capacity, and enables reliable connectivity in diverse operating environments.

4.5 ANTENNA DESIGN CONSTRAINT FOR IOT AND 5G APPLICATIONS

Antenna design for IoT and 5G applications is subject to various constraints that arise from the specific requirements and challenges of these domains. These constraints play a crucial role in determining the practicality, feasibility, and effectiveness of antenna designs. Here are some common antenna design constraints in the context of IoT and 5G [12, 13].

4.5.1 Size and form factor

IoT devices and 5G infrastructure often have limited space available for antenna integration. Antennas must be designed to be compact and low-profile to accommodate the size constraints of these applications. Miniaturization techniques and novel antenna structures are often employed to achieve compact designs without compromising performance.

4.5.2 Frequency range and bandwidth

IoT and 5G systems operate over a wide range of frequencies, including both sub-6 GHz bands and millimeter-wave bands. Antennas need to be designed to cover the desired frequency bands with sufficient bandwidth to support the required data rates and communication protocols. Wideband and multi-band antennas are often necessary to accommodate the frequency range of these applications.

4.5.3 Radiation pattern

The radiation pattern of an antenna must be tailored to meet the specific coverage requirements of IoT and 5G applications. Antennas may need to exhibit omnidirectional radiation patterns for wide coverage, or directional patterns for targeted communication and interference mitigation. Achieving the desired radiation pattern while considering size constraints can be a challenging design constraint.

4.5.4 Multi-technology coexistence

IoT and 5G applications often coexist with various wireless technologies, including Wi-Fi, Bluetooth, Zigbee, and cellular networks. Antennas must be designed to minimize electromagnetic interference and ensure efficient coexistence with these technologies. Techniques such as filtering, shielding, and frequency planning are employed to mitigate interference and maintain reliable communication.

4.5.5 Power consumption

IoT devices, especially those powered by batteries, have strict power consumption limitations to maximize battery life. Antennas must be designed to operate efficiently and consume minimal power while maintaining good performance. Energy-efficient designs, including low-loss materials and optimized radiation efficiency, are crucial to meet power consumption constraints.

4.5.6 Environmental and operating conditions

IoT and 5G systems are deployed in diverse environments, including indoor, outdoor, and harsh industrial environments. Antennas must be designed to withstand environmental factors such as temperature variations, humidity, vibration, and exposure to dust and chemicals. Antenna materials, ruggedization techniques, and environmental sealing are employed to ensure reliability in these operating conditions.

4.5.7 Regulatory compliance

Antenna designs must adhere to regulatory guidelines and standards related to electromagnetic emissions, specific absorption rate (SAR) limits, and frequency allocation. Compliance with regulatory requirements ensures safe and legal operation of IoT and 5G systems. Antenna designs must consider these constraints and ensure compliance with applicable regulations.

These antenna design constraints pose challenges and require careful consideration to develop effective and practical antenna solutions for IoT and 5G applications. Designers and engineers must balance the trade-offs between size, performance, power consumption, and regulatory compliance to meet the unique demands of these domains. To address the design constraints mentioned earlier for IoT and 5G applications, several technical steps can be taken. Here are some approaches to overcome these constraints [14, 15]:

A. Size and Form Factor Constraint: Utilize miniaturization techniques such as fractal geometries, metamaterials, and compact antenna structures. Explore integrated antenna designs, where antennas are embedded within the existing device components or PCB (Printed Circuit Boards). Investigate novel antenna materials and fabrication techniques to achieve compact form factors without compromising performance.

B. Frequency Range and Bandwidth Constraint: Implement wideband or multi-band antenna designs to cover the desired frequency range. Utilize frequency re-configurable or tunable antenna structures to dynamically adapt to different frequency bands. Explore antenna arrays or beamforming techniques to enhance frequency coverage and bandwidth.

C. Radiation Pattern Constraint: Employ advanced antenna design techniques, such as reflectors, directors, and sub-arrays, to shape the radiation pattern according to the desired coverage requirements. Utilize phased array antennas or electronically steerable antennas to achieve beam forming and dynamically adjust the radiation pattern. Implement smart antenna systems that adaptively adjust the radiation pattern based on the environment and communication needs.

D. Multi-Technology Coexistence Constraint: Employ advanced filtering techniques to mitigate interference from coexisting wireless technologies. Explore frequency planning and spectrum allocation strategies to minimize overlap and maximize coexistence. Utilize advanced interference cancellation techniques, such as adaptive beam forming or interference nulling, to improve coexistence performance [16, 17].

E. Power Consumption Constraint: Optimize antenna designs for higher radiation efficiency to reduce power consumption. Explore energy harvesting techniques to utilize ambient energy sources to power the antenna system. Employ intelligent power management algorithms to dynamically adjust antenna performance based on power availability and requirements.

F. Environmental and Operating Conditions Constraint: Select appropriate antenna materials and coatings that can withstand the environmental conditions and provide durability. Perform rigorous testing and simulation under various environmental conditions to ensure reliable performance. Incorporate environmental sensors and feedback mechanisms to monitor and adapt antenna performance to changing conditions.

G. Regulatory Compliance Constraint: Stay updated with the latest regulatory guidelines and standards to ensure compliance during antenna design. Perform extensive electromagnetic emission testing and SAR analysis to meet the regulatory limits. Collaborate with regulatory bodies and seek their guidance early in the design process to ensure compliance.

By implementing these technical steps, antenna designers and engineers can overcome the design constraints and develop efficient, reliable, and compliant antenna solutions for IoT and 5G applications [18].

4.6 ALGORITHM FOR ANTENNA DESIGN

Designing an algorithm for antenna design in IoT and 5G applications involves a complex set of steps and considerations [19]. While it is not feasible to provide a comprehensive algorithm covering all aspects, I can outline a general algorithmic approach that you can follow as a starting point. Here's a high-level algorithm in Python for antenna design in IoT and 5G applications:

Step 1: Define Design Requirements: Specify the requirements for the IoT/5G application, such as frequency range, bandwidth, gain, and form factor constraints.

Step 2: Research and Selection: Explore existing antenna designs and technologies suitable for the application. Select the most

appropriate antenna type based on trade-offs and performance requirements.

Step 3: Simulation and Modeling: Use electromagnetic simulation software to model and simulate the antenna design. Iterate on the design parameters to achieve the desired performance objectives. Optimize the antenna parameters such as dimensions, feed positions, and materials.

Step 4: Prototyping and Fabrication: Create a physical prototype of the antenna based on the simulated design. Choose fabrication techniques like PCB manufacturing, 3D printing, or custom fabrication methods. Pay attention to fabrication tolerances, material properties, and manufacturing constraints.

Step 5: Testing and Validation: Perform tests on the fabricated prototype to evaluate its performance. Measure key parameters like return loss, radiation pattern, gain, efficiency, and impedance matching. Compare the measured results with the simulated data to validate the design.

Step 6: Iterative Optimization: Analyze the test results and identify areas for improvement or adjustment. Iterate the design process by making necessary modifications to the antenna's geometry, dimensions, or materials. Repeat the prototyping, testing, and validation steps to optimize the performance.

Step 7: Compliance and Certification: Ensure the antenna design complies with relevant regulatory standards and guidelines. Collaborate with regulatory bodies or seek certification from authorized testing labs for compliance approval.

Step 8: Integration and Deployment: Integrate the finalized antenna design into the IoT device or 5G infrastructure. Collaborate with system engineers to ensure seamless integration and compatibility. Conduct real-world testing and validation of the integrated system.

4.7 FUTURE DIRECTIONS

Continued research and development are needed to address the evolving requirements of IoT applications. Antenna designs for IoT need to be further optimized for enhanced performance in terms of gain, efficiency, and coverage range. Integration of multiple antennas for diversity and MIMO (Multiple-Input Multiple-Output) techniques can be explored to improve communication reliability and throughput. Antennas for IoT applications may need to support multiple frequency bands and be compatible with various wireless communication standards to ensure interoperability and connectivity. New materials and fabrication techniques can be investigated to achieve miniaturization, flexibility, and integration of antennas with IoT devices. Antenna designs should consider energy harvesting and power-efficient operation to support the low-power requirements of IoT devices. The increasing deployment of IoT in diverse environments, such as smart cities, industrial automation, and healthcare, will pose new challenges for antenna design, including robustness against interference, mobility, and environmental factors. Overall, while significant progress has been made in antenna design for IoT applications, ongoing research and innovation are required to overcome the evolving challenges and enable seamless connectivity in the expanding IoT ecosystem.

4.8 CONCLUSION

This chapter provided a discussion of the future of antenna design for IoT and 5G applications, discussing emerging trends and technologies such as millimeter-wave communications, massive MIMO, and intelligent antennas. It underscored the need for interdisciplinary research collaborations and standardization efforts to drive innovation in antenna design and facilitate the widespread adoption of IoT and 5G technologies.

Overall, this chapter serves as a valuable resource for researchers, engineers, and industry professionals in developing efficient and reliable antenna solutions that meet the evolving demands of IoT and 5G networks. By understanding the design considerations, advancements, challenges, and future perspectives, stakeholders can contribute to the development of robust and high-performance wireless communication systems.

REFERENCES

1. M.A. Khan et al., "A Compact Re-configurable Antenna Design for IoT Applications," *IEEE Internet Things J.*, vol. 8, no. 4, pp. 2801–2812, Feb. 2021. DOI: 10.1109/JIOT.2020.3049357
2. T. Vazquez-Rodriguez et al., "Design and Optimization of an Ultra-Wideband Antenna Array for 5G mm Wave Applications," *IEEE Trans. Antennas Propag.*, 2022. DOI: 10.1109/TAP.2022.3146472
3. N. Singh and Y. Kumar, "Blockchain for 5G Healthcare Applications: Security and privacy solutions," *IET Digital Library*, vol. 1, pp. 1–30, 2021. DOI: 10.1049/PBHE035E
4. M.Z. Hossain et al., "A Novel Design of Compact Re-configurable Antenna for IoT and 5G Applications," *IEEE Access*, vol. 8, pp. 141624–141637, 2020. DOI: 10.1109/ACCESS.2020.3018851
5. S. Verma, and A. Kumar, "Design of a Compact Dual-Band MIMO Antenna for IoT Applications," *Electronics*, vol. 10, no. 9, p. 1082, Aug. 2021. DOI: 10.3390/electronics10091082
6. R.K. Pokharel, S. Song, and K.B. Kim, "Recent Advances in Antenna Designs and their Applications for IoT Systems," *IEEE Access*, vol. 6, pp. 11414–11428, 2018. DOI: 10.1109/ACCESS.2018.2804578

7. Y. Duan, X. Wang, and L. Hanzo, "Smart Antenna Systems for 5G Networks: Taxonomy, Limitations, and Achievable Rates," *IEEE Commun. Surv. Tutor.*, vol. 20, no. 3, pp. 1616–1643, 2018. DOI: 10.1109/COMST.2018.2810079

8. X. Chen, X. Wu, S. Zhang, and S. Xu, "Millimeter-Wave Antennas for 5G Mobile Communications," *IEEE Trans. Antennas Propag.*, vol. 65, no. 12, pp. 6231–6249, 2017. DOI: 10.1109/TAP.2017.2752719

9. F. Liu, C. Liu, and S. Member, "Antenna-in-Package Design for 5G Mobile Communication Systems," *IEEE Trans. Antennas Propag.*, vol. 65, no. 12, pp. 6861–6865, 2017. DOI:10.1109/TAP.2017.2754218

10. N. Gautam, and K.P. Ray, "Wideband and Multi-Band Antennas for Modern Wireless Communications," *IEEE Antennas. Propag. Mag.*, vol. 56, no. 3, pp. 14–31, 2014. DOI: 10.1109/MAP.2014.6930472

11. M.Z. Hossain, M.S. Alam, and M.R. Islam, "Antenna Design for IoT Applications: A Review," *2020 3rd International Conference on Advances in Electrical Engineering (ICAEE)*, Dhaka, Bangladesh, pp. 1–6, 2020. DOI: 10.1109/ICAEE50512.2020.9337921

12. A.A. Abdulgader, N. Hassan, and F. Al-Turjman, "Antenna Design Considerations for 5G Wireless Networks," *2017 International Symposium on Networks, Computers and Communications (ISNCC)*, Marrakech, Morocco, pp. 1–6, 2017. DOI:10.1109/ISNCC.2017.8072113

13. N. Singh, "IOT Enabled Hybrid Model with Learning Ability for E-Health Care Systems", *Measurement: Sensors*, vol. 24, pp. 1–14, Dec. 2022. DOI: 10.1016/j.measen.2022.100567

14. M. Gapeyenko, R. Zviahin, M. Kozhemyakin, and E. Belenkova, "Design Aspects of Antennas for Internet of Things Applications," *2020 International Conference on Engineering and Telecommunication (EnT)*, Moscow, Russia, pp. 1–4, 2020. DOI: 10.1109/ENT52567.2020.9255646

15. M.A.M. Abdalla, A.A.H. Azremi, and M.F.M. Yusof, "Design and Optimization of Antennas for 5G Mobile Communication Systems," *2016 6th IEEE International Conference on Control System, Computing and Engineering (ICCSCE)*, Penang, Malaysia, pp. 70–75, 2016. DOI: 10.1109/ICCSCE.2016.7893474

16. R. Krishan, "Terahertz Band for Wireless Communication—A Review," In *Terahertz Wireless Communication Components and System Technologies*, M.E. Ghzaoui, S. Das, T.R. Lenka, and A. Biswas, Eds., Springer, Singapore, pp. 153–161, 2022.

17. I. Malhotra, and G. Singh, "Terahertz Technology for Biomedical Application," In *Terahertz Antenna Technology for Imaging and Sensing Applications*, Springer, Cham, Switzerland, pp. 235–264, 2021. DOI: 10.1007/978-3-030-68960-5

18. M. Civas, and O.B. Akan "Terahertz Wireless Communications in Space," *ITU J. Futur. Evol. Technol.*, vol. 2, pp. 31–38, 2021.

19. A.L.A.K. Ranaweera, T.S. Pham, H.N. Bui, V. Ngo, and J.-W. Lee "An Active Meta Surface for Field-Localizing Wireless Power Transfer Using Dynamically Re-Configurable Cavities," *Sci. Rep.*, vol. 9, p. 11735, 2019.

Chapter 5

Machine learning-driven antenna design
Optimizing performance and exploring design possibilities

Sandeep Kumar Jain and Pritesh Kumar Jain
Shri Vaishnav Vidyapeeth Vishwavidyalaya, Indore, India

5.1 INTRODUCTION

The field of antenna design has witnessed a remarkable transformation with the advent of machine learning (ML) techniques. ML has emerged as a powerful tool that offers the potential to optimize antenna performance, enhance efficiency, and significantly reduce design time. By leveraging the capabilities of data analysis, pattern recognition, and optimization algorithms, ML-driven approaches have opened up new horizons in antenna design.

In this chapter, we delve into the realm of machine learning-driven antenna design, focusing on the optimization of performance metrics and the exploration of innovative design possibilities. The integration of ML techniques into antenna design processes has revolutionized the way engineers approach and solve complex design problems. By leveraging vast amounts of data, ML algorithms can identify patterns and relationships that may be difficult to perceive through traditional design methodologies. We begin by providing an overview of various ML techniques, such as neural networks, genetic algorithms, and deep learning, and their applications in antenna design. These techniques have shown immense promise in addressing challenges related to antenna performance optimization, including radiation efficiency, gain, bandwidth, and impedance matching.

Furthermore, ML-driven antenna design facilitates the exploration of unconventional and non-intuitive design configurations. By considering a wide range of design parameters and utilizing ML algorithms, engineers can discover novel antenna structures that push the boundaries of traditional designs. This allows for the creation of antennas with improved performance characteristics and unique capabilities tailored to specific applications. The benefits of ML-driven approaches in antenna design are numerous. Firstly, ML techniques enable the rapid optimization of antenna performance by efficiently exploring the design space and identifying optimal parameter settings. This significantly reduces the design time and effort required, leading to faster and more reliable antenna designs.

Secondly, ML algorithms can reveal hidden insights and correlations in complex antenna datasets, leading to improved understanding and

DOI: 10.1201/9781003422440-5

modeling of antenna behavior. This enhanced understanding enables engineers to make informed design decisions and achieve superior antenna performance. Lastly, ML-driven antenna design encourages innovation by enabling the discovery of unconventional and revolutionary antenna configurations. By leveraging the power of ML techniques, engineers can think beyond conventional design constraints and create antennas with unprecedented capabilities.

In conclusion, this chapter aims to provide a comprehensive introduction to the application of ML techniques in antenna design. By optimizing performance metrics and exploring new design possibilities, ML-driven approaches offer significant advantages in terms of performance improvement, design time reduction, and the discovery of innovative design configurations. We hope that this overview will inspire further research and exploration in this exciting and rapidly evolving field of machine learning-driven antenna design.

5.2 LITERATURE REVIEW

In this section we mention some of the existing research papers, these papers collectively showcase the application of ML techniques, including deep learning, genetic algorithms, surrogate modeling, and more, in antenna design optimization, performance enhancement, and design efficiency improvement. They provide valuable insights into the potential of ML in revolutionizing antenna design processes.

"Deep Learning-Driven Design and Optimization of Antennas" [1]: this paper explores the use of deep learning techniques for antenna design and optimization. It investigates the application of convolutional neural networks (CNN) to automate the design process, improve antenna performance, and accelerate the optimization process.

"Application of Machine Learning for Antenna Design Optimization: A Comprehensive Review" [2]: This review paper provides a comprehensive overview of the application of machine learning in antenna design optimization. It discusses various ML techniques, including neural networks, genetic algorithms, and surrogate modeling, and their use in optimizing antenna performance, efficiency, and other design parameters.

"Multi-Objective Design Optimization of Fractal Antennas Using Surrogate Model and Genetic Algorithm"[3]: This paper presents a methodology for optimizing fractal antennas using a combination of surrogate modeling and genetic algorithms. It addresses the multi-objective nature of antenna design problems and demonstrates the effectiveness of the proposed approach in achieving optimal trade-offs between different design objectives.

"Efficient Antenna Design Using Deep Learning and Surrogate Modeling"[4]: This paper proposes a hybrid approach combining deep learning and surrogate modeling for efficient antenna design. It utilizes deep learning techniques to learn the mapping between antenna geometry and performance, and surrogate modeling to enhance the optimization process by approximating the performance surface.

"Evolutionary-Deep Learning Algorithm for the Design of Wideband Antenna Arrays"[5]: This paper introduces an evolutionary-deep learning algorithm for the design of wideband antenna arrays. It combines the power of evolutionary algorithms with deep learning to optimize the performance of wideband antenna arrays, considering multiple design parameters and wideband characteristics.

"Machine Learning-Assisted Design Optimization of Compact Antennas"[6]: This paper presents a machine learning-assisted approach for the design optimization of compact antennas. It demonstrates the use of machine learning techniques, such as support vector regression and neural networks, to assist in the optimization process, reducing design time and improving antenna performance.

5.3 ML TECHNIQUES USED IN ANTENNA DESIGN

ML techniques can also be applied to antenna design to optimize antenna performance, enhance efficiency, and reduce design time. Here are a few ML techniques commonly used in antenna design [6, 7].

5.3.1 Neural networks

Neural networks can be used to model complex relationships between antenna parameters and performance metrics. By training a neural network on a dataset of antenna designs and their corresponding performance, you can create a predictive model that can suggest optimal designs based on desired specifications. It can identify how neural networks can be used to model complex relationships between antenna parameters and performance metrics, and how they can assist in suggesting optimal designs based on desired specifications.

Neural networks are powerful machine learning models that can learn and capture intricate patterns and relationships within data. In the context of antenna design, a neural network can be trained on a dataset of antenna designs, where each design is associated with its corresponding performance metrics. These performance metrics may include parameters such as impedance matching, radiation pattern, gain, bandwidth, and other desired specifications.

The training process involves feeding the neural network with input data representing the antenna parameters, such as dimensions, material properties, and configuration. The network then processes this input through a series of interconnected layers, known as neurons, and learns to extract and transform the input information to produce an output, which in this case would be the predicted performance metrics.

During the training phase, the neural network adjusts its internal weights and biases iteratively to minimize the difference between the predicted performance metrics and the actual performance metrics provided in the training datasets. This process is typically achieved through an optimization algorithm, such as back-propagation, which calculates the gradient of the network's error and updates the weights accordingly.

Once the neural network is trained, it can be used as a predictive model. Given a set of desired specifications for an antenna design [8], the trained neural network can take the input parameters and generate predictions for the corresponding performance metrics. This allows designers to quickly evaluate and assess the expected performance of various antenna configurations without physically building and testing each design.

Furthermore, the neural network can also provide insights into the relationships between different antenna parameters and their impact on performance metrics. Designers can analyze the learned weights and activation within the network to gain a better understanding of the complex relationships between parameters, helping them make informed design decisions.

It's worth noting that the quality of the neural network model relies heavily on the quality and representativeness of the training datasets. A diverse dataset covering a wide range of antenna designs and performance metrics is crucial for training a robust and accurate predictive model.

In short, by training a neural network on a dataset of antenna designs and their performance metrics, designers can leverage the network as a predictive model to suggest optimal designs based on desired specifications. This approach accelerates the design process, reduces costs, and provides valuable insights into the relationships between antenna parameters and performance metrics.

5.3.2 Genetic algorithms

Genetic algorithms are optimization techniques inspired by natural evolution. In the context of antenna design, genetic algorithms can be used to iteratively generate and evaluate antenna designs based on fitness functions that capture performance criteria such as impedance matching, radiation pattern, and gain. Through generations of evolution, the algorithm converges towards optimal or near-optimal designs.

How genetic algorithms can be used as optimization techniques in antenna design and how they iteratively generate and evaluate antenna designs based on fitness functions.

Genetic algorithms (GAs) are optimization algorithms inspired by the principles of natural evolution and genetics. In the context of antenna design, GAs can be employed to iteratively explore and optimize the design space. The algorithm starts with an initial population of antenna designs, which can be randomly generated or based on prior knowledge. Each design in the population is represented by a set of parameters that define its characteristics, such as dimensions, materials, and configuration.

The GA process involves several steps. First, the initial population undergoes evaluation based on fitness functions. Fitness functions quantify the performance of each antenna design based on specific criteria, such as impedance matching, radiation pattern, gain, or any other desired performance metrics. These fitness functions capture the design objectives and constraints, guiding the optimization process.

Next, the GA selects the fittest designs from the current population, favoring those with higher fitness scores. These selected designs serve as the parents for the creation of the next generation. The GA applies genetic operators, such as crossover and mutation, to combine and modify the genetic information (parameters) of the selected designs. Crossover involves exchanging genetic material between two parent designs, while mutation introduces random changes in the genetic information.

The newly generated offspring designs form the next population, which undergoes the same evaluation and selection process. This iterative cycle of evaluation, selection, and genetic operations continues over multiple generations, with the population evolving towards better designs.

Through generations of evolution [9, 10], the GA converges towards optimal or near-optimal designs that exhibit improved performance based on the fitness functions. The algorithm explores the design space, exploiting promising regions and gradually refining the antenna designs. By selecting the fittest designs and applying genetic operations, the GA effectively mimics the principles of natural evolution, enabling it to identify optimal designs in a large and complex design space.

The convergence and effectiveness of the GA depend heavily on the design of appropriate fitness functions that accurately capture the desired performance metrics and objectives. The fitness functions guide the selection process, ensuring that designs with better performance are more likely to be selected as parents for the next generation. Over time, the GA converges towards a population of antenna designs that exhibit desirable performance characteristics based on the defined fitness functions.

In short, genetic algorithms are optimization techniques inspired by natural evolution. In antenna design, GAs iteratively generates and evaluate antenna designs based on fitness functions that capture performance criteria. Through the evolutionary process of selection, genetic operations, and fitness-driven evolution, the GA converges towards optimal or near-optimal designs, offering a powerful approach for antenna optimization.

5.3.3 Finite Element Method (FEM) and machine learning

The Finite Element Method (FEM) is a numerical method used to solve complex electromagnetic problems. ML techniques, such as regression or neural networks, can be employed to create surrogate models that approximate the solutions provided by FEM. This approach can significantly reduce the computational cost of antenna design optimization, making it more efficient.

Utilization of FEM and machine learning techniques to create surrogate models in antenna design optimization. FEM is a numerical method used to solve complex electromagnetic problems, including antenna analysis and design. It involves discretizing the antenna geometry into smaller elements, applying appropriate boundary conditions, and solving Maxwell's equations numerically. FEM provides accurate solutions for antenna performance metrics, such as impedance matching, radiation pattern, and gain.

However, FEM simulations can be computationally expensive, especially for large and complex antenna designs. This is because FEM requires solving a system of equations for each element in the discretized model, which can be both time-consuming and resource-intensive.

To address this issue, ML techniques, such as regression or neural networks, can be employed to create surrogate models. These surrogate models approximate the solutions provided by FEM simulations, reducing the need for repeated and computationally expensive simulations.

Regression models can be trained to learn the relationship between antenna design parameters and performance metrics by using a dataset of FEM simulation results. The regression model can then predict the performance of new antenna designs based on their parameters without the need for actual FEM simulations. This significantly reduces the computational cost, as the regression model provides fast and efficient approximations.

Similarly, neural networks can be trained as surrogate models to approximate the solutions provided by FEM simulations. The neural network takes the antenna design parameters as input and produces predictions for the performance metrics as output. By training the neural network on a dataset of FEM simulation results, it learns the complex relationships between design parameters and performance, enabling it to make accurate predictions.

The surrogate models created using ML techniques serve as efficient alternatives to full-scale FEM simulations. They provide fast evaluations of antenna designs, allowing for quick exploration of the design space and optimization iterations. Surrogate models can be integrated into optimization algorithms such as genetic algorithms or gradient-based methods, enabling more efficient antenna design optimization.

It's important to note that while surrogate models can greatly reduce the computational cost, they are approximations of the true FEM solutions. Therefore, the accuracy of the surrogate models depends on the quality and representativeness of the training datasets. It's necessary to ensure that the datasets cover a wide range of antenna designs and capture the key factors affecting performance.

In short, ML techniques, such as regression or neural networks, can be employed to create surrogate models that approximate the solutions provided by FEM simulations in antenna design optimization. These surrogate models significantly reduce the computational cost by providing fast and efficient approximations. They allow for quick evaluations and exploration of the design space, enabling more efficient and effective antenna design optimization processes.

5.3.4 Reinforcement learning

Reinforcement learning can be applied to antenna design by formulating it as a sequential decision-making problem. The design process is divided into discrete actions (e.g., changing antenna dimensions, adding or removing elements) and an agent learns to take actions that maximize a reward signal, which can be defined based on performance metrics. Reinforcement learning can help explore the design space and find novel antenna configurations.

How reinforcement learning (RL) can be applied to antenna design by formulating it as a sequential decision-making problem and how it can help explore the design space and find novel antenna configurations.

Reinforcement learning is a machine learning technique that focuses on training an agent to make sequential decisions in an environment to maximize a reward signal. In the context of antenna design, RL can be used to optimize antenna configurations by formulating the design process as a sequential decision-making problem.

The design process [6] is divided into discrete actions that the agent can take, such as changing antenna dimensions, adding or removing elements, or modifying other design parameters. These actions affect the antenna's characteristics and ultimately impact its performance metrics, such as impedance matching, radiation pattern, or gain.

The agent learns to take actions by interacting with the environment, which simulates the evaluation of antenna designs based on their performance. At each step, the agent selects an action based on its current state and policy, which is a set of rules or strategies for decision-making. The policy can be deterministic or stochastic, depending on the desired exploration–exploitation trade-off.

After taking an action, the agent receives feedback in the form of a reward signal, which reflects the performance of the antenna design resulting from the action taken. The reward signal can be defined based on the desired performance metrics, with higher rewards given for better performance. For example, an agent could receive a positive reward for achieving good impedance matching or a negative reward for violating specific constraints.

Through a trial-and-error process, the agent learns to adjust its policy to maximize the cumulative reward obtained over multiple steps. Reinforcement learning algorithms, such as Q-learning or policy gradients, are used to update the agent's policy based on the observed rewards and the chosen actions.

By training the agent over multiple iterations, it explores the design space and learns to find novel antenna configurations that maximize the reward signal. RL allows the agent to balance exploration and exploitation, encouraging the discovery of new and promising designs while leveraging known successful designs.

Reinforcement learning can be particularly beneficial when the design space is large and complex, and manual search or optimization methods become challenging or time-consuming. RL provides an automated and adaptive approach to explore the design space efficiently, leading to the discovery of innovative antenna configurations that might not have been considered through traditional design approaches.

It's important to note that RL-based antenna design requires careful formulation of the reward signal and the definition of appropriate state representations to capture the relevant aspects of the antenna's performance [9]. Expert knowledge and domain expertise play a crucial role in guiding the RL process and defining the problem set-up.

In short, by formulating antenna design as a sequential decision-making problem, reinforcement learning can optimize antenna configurations by training an agent to take actions that maximize a reward signal based on performance metrics. RL enables the exploration of the design space, leading to the discovery of novel and innovative antenna configurations that might not have been explored through traditional methods.

5.3.5 Generative adversarial networks (GANs)

Generative Adversarial Networks (GANs) can be used to generate novel antenna designs. By training a GAN on a dataset of existing antenna designs, the generator network learns to create new designs, while the discriminator network evaluates the quality and performance of these generated designs. This iterative process can lead to the generation of unique and innovative antenna designs.

How GANs can be used to generate novel antenna designs and the iterative process involving the generator and discriminator networks.

GANs are a type of machine learning model consisting of two components: a generator network and a discriminator network. In the context of antenna design, GANs can be employed to create new and innovative antenna designs based on existing ones.

To train a GAN for antenna design, a dataset of existing antenna designs is required. This dataset serves as the training data for the GAN. The generator network takes random input noise and generates new antenna designs. The goal of the generator is to generate designs that are similar to the ones in the training datasets.

The discriminator network, on the other hand, evaluates the quality and performance of the generated designs. It takes both the generated designs and real designs from the training datasets as input and distinguishes between them. The discriminator's objective is to correctly identify which designs are real and which ones are generated.

The training process of the GAN involves an iterative adversarial game between the generator and discriminator networks. Initially, the generator produces random antenna designs that are likely far from the real designs. The discriminator is trained to distinguish between the generated and real designs accurately.

As training progresses, the generator learns from the feedback provided by the discriminator. It adjusts its parameters to generate designs that are more similar to the real ones, aiming to fool the discriminator into misclassifying them. Meanwhile, the discriminator improves its ability to discriminate between real and generated designs, becoming more adept at distinguishing the two.

This iterative process of training continues, with the generator and discriminator network improving their performance over time. Through this adversarial game, the GAN converges towards a point where the generated designs closely resemble the real designs in the training datasets [11, 12].

The advantage of using GANs in antenna design is their ability to generate unique and innovative designs that may not have been explicitly present in the original datasets. The generator network learns the underlying patterns and structure of the antenna designs, enabling it to create new designs that exhibit similar characteristics.

GANs can be particularly useful when designers want to explore and generate diverse antenna designs that push the boundaries of traditional design approaches. By leveraging the generative capabilities of GANs, novel and innovative designs can be generated, potentially leading to improvements in antenna performance or the discovery of unconventional design solutions.

It's important to note that the quality of the generated designs depends on the quality and representativeness of the training datasets. A diverse and comprehensive dataset is crucial to ensure that the GAN captures the necessary design principles and characteristics.

In short, GANs can be used to generate novel antenna designs by training a generator network on a dataset of existing designs and having a discriminator network evaluate the quality and performance of the generated designs. The iterative process between the generator and discriminator leads to the generation of unique and innovative antenna designs that may not have been explicitly present in the original datasets. GANs offer a powerful approach for exploring the design space and pushing the boundaries of traditional antenna design.

These ML techniques [13, 14] can assist in antenna design by optimizing performance, reducing design time, and exploring new design possibilities. However, it's important to note that expertise in antenna theory and design principles is still crucial for evaluating and validating the results obtained from ML-based approaches.

5.4 STEPS TO APPLY MACHINE LEARNING TECHNIQUES IN ANTENNA DESIGN

To apply machine learning (ML) techniques in antenna design, you can follow these general steps [4, 5].

5.4.1 Define the problem

Clearly define the objective of your antenna design project. Determine the performance metrics you want to optimize or the specific design challenge you want to address.

5.4.2 Data collection

Gather a relevant dataset of antenna designs, simulation results, and performance measurements. This dataset will be used to train and validate your ML model. Ensure the data is diverse, representative, and accurately labeled.

5.4.3 Pre-process the data

Prepare the collected data for ML analysis. This step may involve cleaning the data, removing outliers, normalizing features, and splitting the dataset into training and testing sets. It is essential to ensure data quality and maintain the integrity of the dataset.

5.4.4 Feature engineering

Identify the relevant features or parameters that influence antenna performance. Extract and engineer these features from the raw data to provide meaningful inputs for your ML model. Consider both geometric parameters (antenna dimensions, materials) and electromagnetic parameters (frequency, impedance, radiation patterns).

5.4.5 Model selection

Choose an appropriate ML algorithm or model architecture for your antenna design problem. This choice may vary, depending upon the specific requirements, the available data, and the complexity of the design task. Commonly used ML models for antenna design include regression models, decision trees, random forests, neural networks, and genetic algorithms.

5.4.6 Model training

Train your chosen ML model using the prepared datasets. During training, the model learns the relationship between the input features and the desired antenna performance outcomes. The model adjusts its internal parameters to minimize the difference between predicted and actual performance values.

5.4.7 Model evaluation

Assess the performance of your trained ML model using the testing dataset. Measure relevant performance metrics, such as prediction accuracy, mean squared error, or correlation coefficients. Evaluate the model's ability to generalize and make accurate predictions on unseen data.

5.4.8 Optimization and design exploration

Utilize the trained ML model to optimize antenna designs. Apply optimization algorithms, such as genetic algorithms or gradient-based methods, to search the design space and find optimal or near-optimal antenna configurations. Explore novel design possibilities by generating new designs based on the learned patterns and rules extracted from the ML model [15].

5.4.9 Model validation

Validate the optimized antenna designs through electromagnetic simulations or physical prototyping. Compare the performance of the optimized designs with existing benchmark designs or desired specifications. Iterate and refine the designs if necessary.

5.4.10 Knowledge transfer and future improvements

Extract knowledge and insights gained from the ML analysis. Document the lessons learned, new design principles, and guidelines derived from the ML-driven antenna design process. Apply this knowledge to future design projects and continuously improve the ML models and methodologies used.

Remember that the effectiveness of ML techniques in antenna design depends on the quality and representativeness of the data, appropriate model selection, and the careful interpretation of results. It may require expertise in both antenna design principles and ML methodologies to ensure successful implementation.

5.5 BENEFITS OF USING ML TECHNIQUES IN ANTENNA DESIGN

Effectiveness of ML techniques in antenna design [16] may depend on various factors such as the availability of relevant data, the complexity of the design problem, and the expertise of the designer in leveraging ML tools effectively [3, 17] (Table 5.1).

Table 5.1 Comparative chart outlining some of the benefits of using ML techniques in antenna design

Benefits	Description
Performance Optimization	ML algorithms can analyze large datasets and optimize antenna parameters to enhance performance metrics such as gain, directivity, bandwidth, and efficiency. This can result in improved antenna designs that meet specific requirements and achieve better performance.
Design Time Reduction	ML algorithms can automate various aspects of antenna design, reducing the time and effort required for manual design iterations. They can rapidly explore design spaces, generate multiple design options, and evaluate their performance, thereby accelerating the design process.
Exploration of New Design Possibilities	ML algorithms can explore unconventional or complex antenna designs that may not be easily envisioned or analyzed using traditional methods. By leveraging large datasets and advanced pattern recognition, ML techniques can discover innovative antenna configurations and push the boundaries of design possibilities.
Pattern Synthesis and Optimization	ML algorithms can synthesize antenna patterns by learning from existing designs or optimization algorithms. They can generate antenna geometries that produce desired radiation patterns, enabling the design of antennas with specific beam-forming, null steering, or beam-shaping characteristics.
Multi-Objective Optimization	ML techniques can handle multi-objective optimization problems in antenna design, where conflicting design goals need to be balanced. By considering multiple performance metrics simultaneously, ML algorithms can find trade-offs and Pareto-optimal solutions, allowing designers to choose from a range of optimized antenna designs based on their priorities.
Knowledge Discovery	ML algorithms can analyze large datasets of antenna designs, performance measurements, and simulation results to extract valuable knowledge and insights. This knowledge can help in understanding the underlying physics of antennas, identifying trends, and guiding future design improvements.

5.6 CONCLUSION

In conclusion, machine learning-driven antenna design represents a compelling and rapidly evolving field with vast opportunities for performance optimization and the exploration of design possibilities. By embracing this approach, antenna designers can overcome design challenges, improve performance metrics, and create innovative antennas that meet the ever-growing demands of modern communication systems.

REFERENCES

1. Zhendong Song, and Jinping Ma. "Deep Learning-driven MIMO: Data Encoding and Processing Mechanism," *Phys. Commun.*, vol. 57, p. 101976, 2023.
2. M.M. Khan, S. Hossain, P. Mozumdar, S. Akter, and R.H. Ashique. "A Review on Machine Learning and Deep Learning for Various Antenna Design Applications," *Heliyon.* vol. 8, no. 4, 2022.
3. Jian Dong, Wenwen Qin, and Meng Wang. "Fast Multi-Objective Optimization of Multi-Parameter Antenna Structures Based on Improved BPNN Surrogate Model," *IEEE Access*, vol. 7, pp. 77692–77701, 2019.
4. Z. Qing, X. Qing, S. M. Ali, et al., "Efficient Antenna Design Using Deep Learning and Surrogate Modeling," *IEEE International Symposium on Antennas and Propagation (APSURSI)*, 2019.
5. Chao Wu, et al. "Evolving Deep Convolutional Neutral Network by Hybrid Sine–Cosine and Extreme Learning Machine for Real-Time COVID19 Diagnosis from X-ray Images," *Soft Comput.*, vol. 27, no. 6, pp. 3307–3326, 2023.
6. M.O. Akinsolu, K.K. Mistry, B. Liu, P.I. Lazaridis, and P. Excell. "Machine Learning-assisted Antenna Design optimization: A Review and the State-of-the-art," In *2020 14th European conference on antennas and propagation (EuCAP)* 2020 Mar 15 (pp. 1–5). IEEE.
7. N. Singh and Y. Kumar, "Blockchain for 5G Healthcare Applications: Security and privacy solutions," *IET Digital Library*, pp. 1–30, 2021. DOI: 10.1049/PBHE035E
8. A. Reineix, and B. Jecko, "Analysis of Microstrip Patch Antennas Using Finite Difference Time Domain Method," *IEEE Trans. Antenna Propag.*, vol. 37, pp. 1361–1369, 1989.
9. Z. Lou, and J.M. Jin, "Modeling and Simulation of Broad-Band Antennas Using the Time-Domain Finite Element Method," *IEEE Trans. Antenna Propag.*, vol. 53, pp. 4099–4110, 2005.
10. T.K. Sarkar, A.R. Djordjevic, and B.M. Kolundzija, "Method of Moments Applied to Antennas," *Handbook of Antennas in Wireless Communications*, pp. 239–279, 2000.
11. S. Li, Z. Liu, S. Fu, Y. Wang, and F. Xu, "Intelligent Beamforming via Physics-Inspired Neural Networks on Programmable Metasurface," *IEEE Trans. Antennas Propag.*, vol. 70, no. 6, pp. 4589–4599, Jun. 2022, DOI: 10.1109/TAP.2022.3140891
12. Z. Ma, K. Xu, R. Song, C. Wang, and X. Cheng, "Learning-Based Fast Electromagnetic Scattering Solver Through Generative Adversarial Network," *IEEE Trans. Antennas Propag.*, vol. 69, no. 4, pp. 2194–2208, Apr. 2021, DOI: 10.1109/TAP.2020.3026447
13. Z. Wei, and X. Chen, "Physics-Inspired Convolutional Neural Network for Solving Full-Wave Inverse Scattering Problems," *IEEE Trans. Antennas Propag.*, vol. 67, no. 9, pp. 6138–6148, Sep. 2019, DOI: 10.1109/TAP.2019.2922779
14. M. Li et al., "Machine Learning in Electromagnetic with Applications to Biomedical Imaging: A Review," *IEEE Antennas Propag. Mag.*, vol. 63, no. 3, pp. 39–51, Jun. 2021, DOI: 10.1109/MAP.2020.3043469

15. A. Al-Saffar, A. Zamani, A. Stancombe, and A. Abbosh, "Operational Learning-Based Boundary Estimation in Electromagnetic Medical Imaging," *IEEE Trans. Antennas Propag.*, vol. 70, no. 3, pp. 2234–2245, Mar. 2022, DOI: 10.1109/TAP.2021.3111516

16. W. Shao, and Y. Du, "Microwave Imaging by Deep Learning Network: Feasibility and Training Method," *IEEE Trans. Antennas Propag.*, vol. 68, no. 7, pp. 5626–5635, Jul. 2020, DOI: 10.1109/TAP.2020.2978952

17. S. Ledesma, J. Ruiz-Pinales, and M. Garcia-Hernandez, et al., "A Hybrid Method to Design Wire Antennas: Design and Optimization of Antennas Using Artificial Intelligence," *IEEE Antennas Propag Mag.*, vol. 57, pp. 23–31, 2015.

Chapter 6

Antenna design exploration and optimization using machine learning

R Sreelakshmy

Vel Tech Rangarajan Dr Sagunthala R&D Institute of Science and Technology, Chennai, India

V Vishnupriya

Electronics Department, Govt. Model Engineering College, Thrikakkara, Kochi, India

KV Shahnaz

Vel Tech Rangarajan Dr Sagunthala R&D Institute of Science and Technology, Chennai, India

G Dhanya

MGM College of Engineering and Technology, Pampakuda, India

Salini Abraham

Jaibharath College of Management & Engineering Technology, Perumbavoor, India

6.1 INTRODUCTION

Antenna engineers have become interested in machine learning (ML) over the few several years. In most cases, the antenna design procedure necessitates observing current distributions through simulations concerning the realization of the electromagnetic (EM) properties of the antenna. The optimization of the settings then makes use of these EM features. An antenna may be designed using simulations and ML. In the realm of designing antennas, the use of artificial intelligence (AI) can provide positive outcomes. In recent years, the use of evolutionary algorithms (EAs) for antenna synthesis or design optimization has become increasingly popular. Thetwo principal methods used for antenna synthesis at the moment are particle swarm optimization (PSO) and differential evolution (DE).

When an element is located, the smart antenna array may be reconfigured to fulfill the goal of elegantly degrading the beam-forming and beam-steering performance. It can be accomplished by simply rearranging the array when a faulty element is discovered. Through optimization, Support Vector Machines

DOI: 10.1201/9781003422440-6

(SVM) and ML may be used to achieve this reconfiguration. The classifica-tion and regression problems can be solved by a special type of supervised learning algorithm called SVM. They provide excellent performance in gen-eralization ability and complexity computation and are therefore capable of solving problems associated with antenna array processing such as beam forming and the estimation of angle of arrival. The prime idea is to adapt the excitation coefficient for each array member to compensate for changes brought on by the environment (magnitude and phase). In order to retain a certain radiation pattern, improve its beam-steering and nulling qualities, and address the DOA problem, the antenna array is trained to adapt its con-stituent parts using SVM [1] based on optimization. As a useful tool in a variety of antenna designs and associated applications, ML is one of the most promising and prominent study topics in AI.

In order to optimize the antenna performances, local and/or global numerical optimization techniques are mostly utilized. Although experience-driven parameter sweeping is clearly superior to numerical optimization, there are still certain restrictions [2]. This technique might be time-consuming for contemporary antennas with several sensitive parameters, with no assur-ance of satisfactory results – it is typically a matter of trial and error. Numerous electromagnetic (EM) simulations were conducted in order to identify almost perfect designs. For a detailed characterization of antennas, numerical methods based full-wave EM simulations are required. Full-wave EM simulations are computationally very upscale by definition. There is no problem with using one EM simulation to describe a design; however, the several EM simulations demanded by global optimization methods impose an unsupportable strain on computers [3].

6.2 MACHINE LEARNING

Machine learning (ML) incorporates a variety of computer methods in order to enhance the performance by automating the data acquisition. Being one of the data-driven optimization methods, ML implements regression meth-odology in order to provide higher levels of automation, progressively in the restoration of time-consuming manual tasks, with automated methods that increase certainty by recognizing and exploiting regularities in sample data [4]. Figure 6.1 shows the primary three categories of machine learning. In actuality, many data-driven and machine learning (ML) approaches, such as conventional artificial neural networks (ANNs), a few decades ago, were initially created and studied in the field of electromagnetics. However, these previous studies did not benefit from the most recent improvements in ML, which have been utilized by the current convergence of smarter network algorithms and architectures, data science, and improved hardware per-formance at lower prices [5]. The computational electromagnetics (CEM) benchmark has now undergone considerable study.

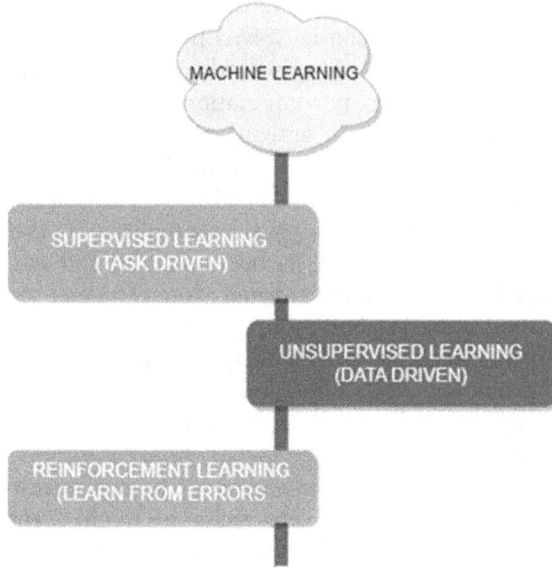

Figure 6.1 Machine learning.

A wider family of ML techniques built on ANNs has been developed today. Deep reinforcement learning, generative adversarial networks, recurrent neural networks, convolutional neural networks, and deep neural networks are just a few of these examples [6]. These methods have been employed successfully to address a number of technical and scientific issues, including, but not limited to, issues with autonomous cars, social media platforms, virtual personal assistants, and picture and video recognition. This normally suggests that using machine learning to genuine electromagnetic problems is one of the ML and AI advances that may be forthcoming. In fact, ML has become a crucial component of electromagnetics' experimental, computational, and theoretical aspects.

Figure 6.1 depicts how the design of antenna arrays and artificial electromagnetic media, such as metamaterials, metasurfaces, electromagnetic band gap structures, and frequency selective surfaces, benefits greatly from the wealth of techniques provided by ML, which can be used to extract the best geometric and structural patterns from highly dimensional stochastic data [7].

In the near future, it is anticipated that ML tools will form the centerpiece of antenna designs. The use of conventional analytical and numerical techniques for the design of complex antenna arrays and electromagnetic structures is very difficult. It has been demonstrated that novel ML methods have the capability to become powerful optimization tools and enablers for these designs by utilizing randomized numerical and experimental data. Modern antenna and metamaterial designs have several objectives spanning the temporal, frequency, and spectrum domains as well as numerous limitations, which presents major optimization challenges.

Due to the fact that these electromagnetic issues typically take on nonlinear and multi-scale shapes, resulting in high-dimensional and non-convex optimization landscapes, and substantial mutual connection (coupling). Fortunately, ML has made it possible to solve these once-unsolvable optimization issues quickly and effectively. Given that ML tools are often simple to use and are highly tempting to antenna engineers since they do not require specialized expertise; nonetheless, others may view this as a possible drawback [7]. Although these algorithms have a lot of promise, it's vital to stress that they must be utilized correctly and that there is no one single tool that can handle all problems, as was discovered from the papers gathered related to this.

When using electromagnetic modeling tools like ML, data-driven design and optimization appropriate previous knowledge, and validation must be required. The gathered papers could serve as a manual for crucial ML practice and the particulars of usability [8].

In recent years there have been intriguing similarities betweenthe emergence of ML and high-performance computing in the field of CEM [5]. Both methods offer strong tools for analyzing complex electromagnetic propagation and scattering problems and gaining in-depth insights into electromagnetic physics, allowing the community to address scientific and engineering issues on a scale and with a scope never before possible. The use of CEM techniques accompanied by ML can provide yet another useful viewpoint to more conventional methods, improving the efficiency and precision of current CEM algorithms. In the case of scattering, inverse scattering, imaging applications, radar signal processing, and reinforcement strategies for CEM, the successful candidates are AI and ML methods [6].

New generations of antenna and radar systems can be developed using the combination of ML algorithms, measurement data, and numerical modeling. They also help in the development of ultra high-performance and multidimensional CEM solvers. It may be persuaded by the findings in it that ML and data-intensive analysis may widen the area of study by complementing other well-known CEM methodologies.

The main aim of this special issue is to provide many fascinating lines of recent research in the field of AI that mix machine learning with data-driven methodologies, antennas, and radio waves. In order to boost their visibility within the electromagnetic community and to solve scattering and propagation concerns, it especially seeks to present a wider viewpoint by emphasizing current possibilities, triumphs, and challenges for machine learning (ML) and their promising applications to antennas, radar, and other devices [9].

Two PIN diodes have been utilized to switch the antenna, which has a square configuration. A square-shaped planar monopole antenna is created for WLAN, Wi-MAX, and Wi-Fi applications. The idea of a truncated ground plane is applied here. The antenna is made from a FR4 substrate, which has a thickness of 1.6 mm. Truncated metallic ground is utilized to get the best radiation pattern and radiation efficiency. This antenna has not satisfied the GPS frequency range and is larger than the antenna described in this study.

Various geometrical forms have been employed to provide frequency reconfigurability. It is possible to accomplish frequency reconfiguration by using fractal slots. For various wireless applications, the multiband frequency reconfigurable antenna is suggestible. A hexaband frequency reconfigurable antenna that can support a variety of applications, including UMTS, WiFi, Bluetooth, WiMAX, and WLAN, has been suggested [10].

The majority of reference antennas do not support GPS applications, and their diameters are significantly greater than those of the proposed antenna. Using Ansys software, a small, frequency-reconfigurable monopole antenna has been created, which has been modified for various bands. This section includes explanations of the switching method, planned antenna shape, and antenna design theory. The antenna construction has employed a total of two switches to accomplish frequency reconfigurability. The PIN diode's lumped equivalent model has been employed in the simulation for both ON and OFF situations [11].

6.3 ANTENNA

An antenna can be used for transmission or reception. An antenna that transmits electromagnetic waves that are produced by electrical impulses is known as a transmitting antenna. Conversely, an antenna that receives electromagnetic waves and converts them into electrical impulses is referred to as a receiving antenna. Finally, when used as transceiver, the same antenna act as both transmitter and receiver. Antennas are sometimes referred to as aerials. They have antennas, either a single one or multiple ones. Over time, antennas have seen substantial changes in terms of both size and appearance.

There are different types of antennas depending on their intended purposes, including loop antenna, horn antenna, patch antenna etc. Wherever a need for wireless communication arises, an antenna can be utilized. When installing a wire system is not practical, electromagnetic waves can be transmitted or received for communication purposes using an antenna. There are various reasons why we use or need antennas, but one particularly one is that they provide an easy way to transmit a signal (or data) when other methods are difficult. Take the case of an airplane, for example. The pilot must communicate frequently with the Air Traffic Control (ATC) personnel. It would be insane to connect the ATC with a cable that is connected to the plane's tail and has a dynamically adjustable length. Wireless communication is the only feasible option, and antennas are where it starts [12].

6.3.1 Microstrip patch antenna

A microstrip or patch antenna emits electromagnetic waves when power from a feed line hits the antenna's strip. The patch's waves radiate outward, starting on the width side. The strip's thinness causes the edge to function as a reflector for the waves produced within the substrate, however. The strip's continuous structure along its length prevents radiation from being released [6].

Additionally, when the creation of the patch endsabruptly, radiations are once again discharged from the patch's second width side. Only a small fraction of the incomingenergyissent off by the patch antenna because the discontinuous structure favours reflections. Only short wave transmission lengths, such as those between neighbouring offices, stores, or other indoor sites, can be covered by microstrip antennas dueto theirrelatively poor radiatingcapacity [13].

It is common practice to provide hemispheric coverage at an angle of 30 to 180 degrees width from the mount using a patch antenna. The ground plane must be much larger than the antenna patch in size. On the substrate, photo-etching is used to create the radiating element and feed lines. The antenna patch has to be a tiny, conductive patch. A thick dielectric substrate with a dielectric constant between 2.2 and 12 will produce good antenna performance. Microstrip component arrays provide greater directivity in the antenna configuration. High beamwidth is provided via low-profile antennas. A very important element is a patch antenna provides. Low-efficiency and a limited bandwidth are the effects of high Q. The substrate's thickness can be increased to compensate for this, though. Beyond a certain point, however, the thickness will result in an unintended loss of power [5].

6.4 SOFTWARE DESCRIPTION

6.4.1 Python

Among the most commonly used programming languages globally, Python is used in everything from machine learning to software testing to websites. However, it isn't specifically designed to design any special part of an application. Red Monk conducted a survey of developers and rated it as the second prominent programming language among them because of its versatility. Python is often used to build apps and websites, automate time-consuming operations, and analyze and display data.

It is a simple, uncomplicated language. Reading well-written Python software is something like reading well-written English. One of Python's most useful features is the pseudo-code architecture. This permits you to concentrate more on the answer to the problem and less on the language itself. Learning Python is very easy. Python's syntax is really simple, as has already been said. Python is an illustration of FLOSS (Free/Library and Open Source Software). This software is an open source software, ie, its source code can be viewed, edited, and some parts of it can be incorporated into other free programs. The sharing of knowledge is centralized to FLOSS. One of its main advantages is that Python was created, and is being improved by, a community that simply wants to improve the language [14].

Python takes care of all the low-level details, such as managing the memory that your software consumes, etc. Python has been modified to run on a variety of platforms thanks to its open-source nature. All of your Python applications will operate on any of these platforms without the need for any modifications if you are careful enough to avoid any system-dependent features. Many other operating systems, such as GNU/Linux, Windows, FreeBSD, Macintosh, Solaris, OS/2, Amiga, AROS, AS/400, BeOS, OS/390, and coding: utf-8z/OS, Palm OS, are also compatible with Python. The original code is internally converted by Python into byte codes, an intermediate format that is subsequently converted into the computer's native language and executed. Python is really a lot easier to use since you don't have to worry about constructing the software or making sure the correct libraries are linked and loaded, etc. Moreover, this substantially improves the portability of your Python applications, since you can simply copy them to another computer and run them there [10].

The reusable sections of code in procedure-oriented languages are procedures and functions. Data and functionality are combined in objects that are written in object-oriented languages [15]. Python offers a highly powerful yet simple object-oriented programming (OOP) approach, especially when compared to more powerful languages like C++ or Java. A C or C++ component can be included in your Python script if you require a critical piece of code to run rapidly or if you wish to close a particular process. Python may be included into your C/C++ projects to provide users access to scripting features [12].

6.4.2 CST studio suite

SIMULIA designed the CST Studio Suite software package for electromagnetic analysis. By using CST Studio Suite, the conduct of ample range of electromagnetic systems can be simulated, for bothlow-frequency and high-frequency applications. Numerous industries are served by CST Studio Suite antenna design. Antennas have been used for many years in the public broadcasting (TV/radio), military, aerospace, and maritime industries. In recent years, antennas have become increasingly prevalent in portable, smart gadgets and appliances in our homes. These gadgets frequently have a variety of antennas for communicating at various frequencies. The most prominent illustration is a smartphone. Antennas are needed in a smartphone for the normal call and internet access on the 4G or 5G mobile networks. For extras such as speakers or headphones, we have a Bluetooth antenna. There includes a Wi-Fi antenna for in-home or workplace internet access, and an NFC antenna enabling mobile payments with a tap of the phone. Many more businesses from all sectors are now making use of electromagnetic simulation to evaluate the performance of the antennas needed to communicate with the rest of the world as a result of our "smart" or "connected" products [16].

For the development, analysis, and optimization of electromagnetic (EM) systems and components, CST Studio Suite®, a high-performance 3D EM analysis software package, is employed. CST Studio Suite provides electromagnetic field solutions in a single user interface for applications for the complete the EM spectrum. The solvers' ability to be coupled to run different types of simulations gives engineers the opportunity to swiftly and efficiently examine whole systems made up of several components.

Due to code sign with other SIMULIA technologies, EM simulation can be incorporated into the design flow and drives the validation process from the start. Antenna and filter performance, electromagnetic compatibility and interference, and the exposure of common EM analysis themes include the response of the human body to EM fields, the electromechanical consequences in motors and generators, and others. And thermal effects in high-power devices. Leading engineering and technology companies use CST Studio Suite all over the world. It facilitatesshorter product development cycles and cheaper development costs, offering significant benefits in terms of getting a product to market. Through simulation, virtual prototyping may be used. Fewer physical prototypes are required, the device's performance can be improved, any errors may be identified and handled early in the design process, and there may be a reduced chance of test failures and recalls [17].

6.5 APPLICATIONS

In several applications, including UAV, THz, textile, GPS, and satellite, machine learning and deep learning are demonstrating excellent outcomes [11]. Due to its effectiveness in a variety of applications in engineering science, medicine, and economics, machine learning has drawn a whole lot of interest. In electromagnetics, the new technique has also been gaining acceptance.

6.6 METHODOLOGY

6.6.1 Design of a microstrip patch antenna

The microstrip patch antenna has grown to be quite well-known and has invited a lot of interest in the research because it is chosen since it is lightweight, compact, affordable, and can retain great performance over a wide variety of frequencies. Using CST Microwave Studio simulation software, the rectangular patch is constructed in this study with many parameters, including return loss, VSWR, directivity in two directions, radiation pattern in 2-D and 3-D, smith chart, and impedance matching. The bandwidth and return loss are both increased by the microstrip patch antenna. For the suggested antenna, FR-4 with a dielectric constant of 4.3 is employed as the substrate. For many applications in the industrial, scientific, and medical sectors, a microstrip rectangular patch antenna with an inset type feed is highly helpful.

A tiny metallic patch is placed on top of a thick metallic ground layer to create the microstrip patch antenna. The substrate, a dielectric sheet, supports the patch. The patch is typically etched onto the dielectric substrate using printed circuit board (PCB) technology. Because of this, a microstrip patch is often known as a printed antenna. The size and shape of a patch affect its performance. The feed for the microstrip patch can be either coaxial type or microstrip type. Etching of both the patch and the microstrip can be done simultaneously.

A correct impedance point can be obtained on the patch by forming a recess on the patch. The impedance is used to calibrate the recess depth. A patch antenna's input bandwidth is fed via a coaxial transmission line of approximately 2 to 4 percent. A single patch antenna has a maximum directional gain of 6–9 db.

The diversity of polarization that a patch antenna may have is one of its inherent advantages. Patch antennas may be constructed with perfect polarization using a single-feed asymmetric patch configurations. By feeding the patch to symmetry around the centre line, unwanted modes are less likely to be excited. Feeding lines will have features of patch impedance for increased performance.

The patch antenna was supplied by a microstrip line, connected to the patch's internal point, when the input impedance is 50 ohms. The most popular type of microstrip patch antenna is rectangular in shape, while the patch can also be square, circular, triangular, orelliptical. Figure 6.2 shows the antenna basic shape with W standing for width and L for length.

6.6.2 Design parameters for patch antenna

The bandwidth and return loss are both increased by the microstrip patch antenna. Considering the substrate, flame-resistant FR-4 with dielectric constant 4.3 is used. For many applications in the industrial, scientific, and medical sectors, a low-profile antenna with rectangle patch using an inset feed approach is considered very helpful. The values of the antenna size may be simply calculated by putting $c = 3108$ m/s, $r = 4.3$, and $f0 = 3$ GHz. Model theory has been applied to determine how long the antenna's radiating element should be. For various resonance frequencies, the computed length of

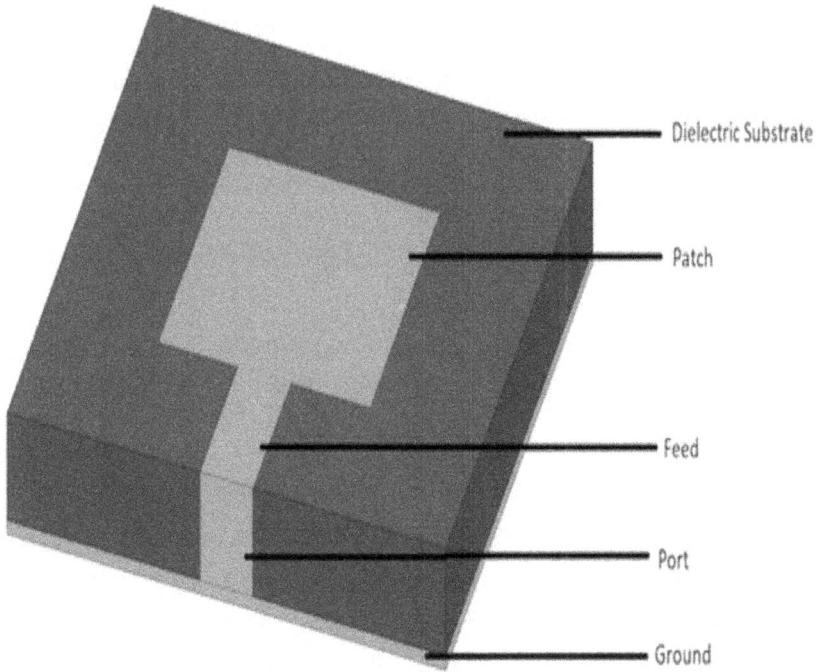

Figure 6.2 Structure of a microstrip patch antenna.

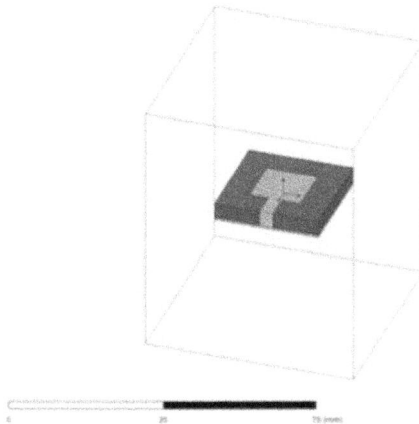

Figure 6.3 Structure of a microstrip patch antenna.

the resonant elements is about equal to a quarter of the wavelength. A slit has been carved at the antenna's ground plane to improve radiation efficiency and the reflection coefficient, as illustrated in Figure 6.3. The FR-4 substrate was utilized for the design because it is inexpensive and widely accessible.

It has a loss tangent equal to 0.02 and relative permittivity equal to 4.4. The antenna is excited by a 50 microstrip line and its width is determined by the general equation. A square patch antenna with the following characteristics: resonance frequency ($f0$), length (L), and patch width (W) [6].

6.6.3 Simulation

CST is used to model and develop the microstrip patch antenna. Figure 6.4 shows the antenna S parameters to represent the reflection coefficient. This program does a broad analysis of 3D and multilayer topologies. It has frequently been utilized in the design of several antenna designs. It may be used to figure out and plot radiation patterns, E-field and H-field distribution, gain, real power vs frequency, standing wave ratio from Smith charts, and return loss [18]. A CST Studio Suite may be used to design, optimize, and analyze various antenna types for a variety of purposes. This program allows for the calculation and optimization of all antenna characteristics, including resonant frequency, impedance matching, bandwidth, gain, beam width, and polarization. Table 6.1 shows the various design parameters used in the simulation.

Before moving forward with physical prototypes, electromagnetic simulation helps to increase the efficiency of the product, verify compatibility, guarantee the product's performance in a complicated and loud EM environment, and also assess the dependability and safety of the goods. We can see the surface current distribution on the antenna and the invisible electromagnetic wave propagation with the help of CST Studio Suite. Designers

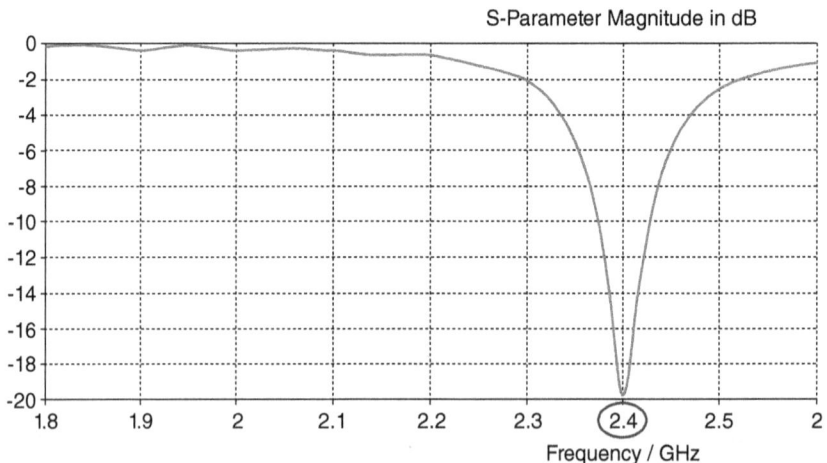

Figure 6.4 S parameter.

Table 6.1 Design parameters

Parameter	Value
Operating frequency (f0)	2.4 GHz
Dielectric substrate	FR-4
Dielectric constant of substrate (er)	4.3
Height of the dielectric substrate (h)	1.6 mm
Return Loss, S11	−10 dB
Voltage Standing Wave Ratio (VSWR)	0.035 mm
Perfect Electric Conductor (PEC)	0.035 mm
Impedance of the Antenna	50 Ω

may check for failures with the aid of CST Studio Suite, but they can also identify its root cause at an early stage [19].

6.6.4 Selection and loading

Selection of the correct data source, data type, and collection tools is known as data selection [12]. Data relevant to the analysis are chosen and obtained from the data-gathering procedure during data selection, which comes before the real practice of data collection.

6.6.5 Data preprocessing

Data can be largely useless and incomplete. To check this, do a data clean-up. Missing data creates a gap when dealing with missing or noisy data [20]. There are many ways to fix this. Bundles are ignored. This method is only suitable when a tuple has many missing values and the volume of data we have is large.

This work can be done in many ways to complete the missing element. Missing values can be filled in manually by the character definition or possibly values. Encode data by category. Because categorical data is a separate variable, most machine learning algorithms require an input number and a label for output. Convert categorical data to numeric data using a number and gold. Figure 6.5 illustrates a flow diagram for data pre-processing.

6.6.6 Splitting data in to train data and test data

Data splitting is the process of splitting the currently accessible data into two sections, usually for cross-validator needs [21]. A component of the data is used to build an equivalent model, and a distinct portion of the information

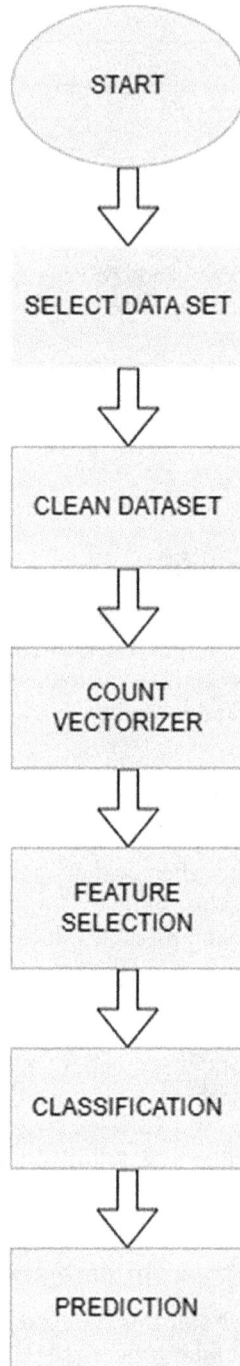

Figure 6.5 Flow diagram.

is adopted to determine the model's efficiency. Building training and testing sets of data is a demanding stage in the analysis of data mining algorithms. When the information set is divided into a training set and a testing set, the majority of the data is often utilized for training and a limited set for testing [22]. Regardless of the type of dataset being used, the training and testing data is divided to train any machine learning model. The problem of categorizing a new observation is to determine which of the set of categories it belongs to, based on a training set of data that comprises observations and whose category membership is known.

One of the most effective machine learning algorithms is the SVM. Given a small sample size, it will strike a compromise among the model complexity and classification capability. The SVM has numerous advantages over other machine learning techniques in that it may function without any prior information and can avoid the effects of noise. An input to one of two classes is predicted by the SVM, a non-probabilistic binary linear classifier, for every input.

It improves the classification and linear analysis of hyperplane creation methods. In machine learning, classification and regression are the two major applications of the neural network (NN) algorithm. To establish the category of an unknown sample, all the training dates are used as reference points and then the NN is used to calculate the distances between the unknown sample and all the training sample points. The category serves as the only criteria for choosing the unidentified sample category. The K-nearest Neighbor (KNN) approach is devised because the NN algorithm is highly sensitive to noise input [23]. The key idea behind the KNN is that after knowing data and tags from the training, the test data is fed, their features are compared to those from the training set, and the K with the highest similarity is discovered.

6.6.7 Prediction

Predictive analytics systems try to obtain the minimum error possible either by "boosting" or by "bagging". There are certain definitions related to this. The accuracy is the skill of the classifier. It defines the class label exactly, and predictor accuracy describes the effectiveness of a particular predictor while making an educated estimate on the value of a predicted characteristic for fresh data [24]. Speed is the measure of how quickly a classifier or predictor can be created and used. Robustness describes a classifier's or predictor's capacity to provide very exact predictions from the supplied noisy data. Scalability is the capacity to efficiently build the classifier or predictor provided a huge amount of data interpretability [25]. It speaks to the level of comprehension of the classifier or predictor.

6.7 CONCLUSION

The aforementioned conversation has shone some light on the significance and utility of using ML approaches to the design and evaluation of several antennas. However, when using this strategy as opposed to computational electromagnetic, various difficulties appear. The absence of standardized data sets for antenna structures that can be utilized directly to train any particular model and give results is the first problem. Instead, data must first be produced through simulations in order to build a data base of chosen input and output variables. Since the primary objective of employing ML in the area of antennas is to gain a quicker design and optimization process while maintaining a high level of accuracy, this might be a laborious and time-consuming effort. A greater computational burden results from the need to run simulations in order to collect a data set. Choosing the optimum model hyperparameters that can produce the greatest outcomes is another factor to take into account. It is obvious that ANNs have dominated this research field because they are the most popular ML technique, have a wide range of frameworks and software packages at their disposal for their fast and effective employment, and have proven resilient in producing results that are highly accurate when compared to traditional CEM approaches. The significance of ANNs in antenna design becomes increasingly apparent as the level of antenna structure complexity rises. The best training and optimization methodologies, network design, regularization methods, activation function selection, and other parameters that impact the performance of the model must thus be examined for each kind of antenna. Although ML emerges as an appealing antenna design and analysis tool that can provide predictions with a high degree of accuracy in a limited amount of time than simulation techniques, needing to create the training data would appear unappealing and difficult. The creation of an antenna design software that is solely based on ML models to change simulators is thus a promising strategy for resolving this issue. The designer flexibility of such a tool would undoubtedly be constrained, and it would be focused on particular antenna kinds and structures; however, it may be expanded to encompass a huge number of antennas. The rapid characterization of the chosen antenna type would be possible with the aid of a rapid, accurate, and optimized design tool. The design requirements are simply entered by the user to get the geometrical predictions. However, when a unique structure is required, this programme falls short, necessitating simulations on the part of the designer. This work included a thorough use of ML to antenna design and analysis. Simulators place a lot of strain on computers, and thus ML is intended to lighten that load and speed up design. This chapter has considered which research articles have used ML algorithms in their design across the literature. Additionally, a summary of several ML principles has been provided, giving readers who are interested in antenna research, but who are unfamiliar with ML, the core knowledge they need to employ these powerful tools in their projects.

REFERENCES

1. T. Gayatri, G. Srinivasu, D. Chaitanya, and V. Sharma "A Review on Optimization Techniques of Antennas Using AI and ML/DL Algorithms," *Int. J. Adv. Microw. Technol*, vol. 7, no. 2, pp. 288–295, 2022.
2. Abdelaziz A. Abdelhamid, and Sultan R. Alotaibi, "Deep Investigation of Machine Learning Techniques for Optimizing the Parameters of Microstrip Antennas," *Int. J. Eng. Technol. Manag. Appl. Sci.*, vol. 12, no. 13, 12A13N, pp. 1–15, 2021.
3. H.M. El Misilmani, N. Tarek, S.K. Al Khatib, "A Review on the Design and Optimization of Antennas Using Machine Learning Algorithms and Techniques," *Int. J. RF Microw. Comput.-Aided Eng.*, vol. 30, pp. 1–28, 2020.
4. Q. Wu, H. Wang, and W. Hong, "Broadband Millimeter-Wave SIW Cavity-Backed Slot Antenna for 5G Applications Using Machine-Learning-Assisted Optimization Method," *International Workshop on Antenna Technology (iWAT) Miami, FL*, 2019.
5. Rajkamal K., and Immadi G, "Optimization of UWB Array Antenna for Bio-Medical Applications Using Cuckoo Search Algorithm," *Int. J. Eng. Technol.*, vol. 7, pp. 367–383, 2018.
6. K.C. Sri Kavya, S.K. Kotamraju, M. Sreevalli, M.P. Priyamvada, Teja K. Ravi, and M. Asma, "Design and Optimization of Array Antennas Using Artificial Intelligence," *Int. J. Eng. Innov. Technol.*, vol. 8, pp. 921–926, 2019.
7. N. Singh and Y. Kumar, "Blockchain for 5G Healthcare Applications: Security and privacy solutions," *IET Digital Library*, pp. 1–30, 2021. DOI:10.1049/PBHE035E
8. R. Jiang, X. Wang, S. Cao, J. Zhao, and X. Li, "Deep Neural Networks for Channel Estimation in Underwater Acoustic OFDM Systems," *IEEEAccess*, vol. 7, pp. 23579–23594, 2019.
9. M.D. Gregory, Z. Bayraktar, and D.H. Werner, "Fast Optimization of Electro-magnetic Design Problems Using the Covariance Matrix Adaptation Evolutionary Strategy," *IEEE Trans. Antennas Propag.*, vol. 59, no. 4, pp. 1275–1285, 2011.
10. M. Chen, U. Challita, W. Saad, C. Yin, and M. Debbah, "Machine Learning for Wireless Networks With Artificial Intelligence: A Tutorial on Neural Networks," *CoRR*, vol. abs/1710.02913, 2017. https://infinite-vigilant.github.io/AI.pdf
11. N. Singh, G. Kumar Prajapati, "Wireless Sensor Network Security Problems-A Review", *International Journal of Emerging Technology and Advanced Engineering*, vol. 8, no. 1, pp. 77–80, 2018.
12. Z.M. Fadlullah, M.M. Fouda, N. Kato, X. Shen, and Y. Nozaki, "An Early Warning System Against Malicious Activities for Smart Grid Communications," *IEEE Netw.*, vol. 25, no. 5, pp. 50–55, 2018.
13. X.H. Chen, X.X. Guo, J.M. Pei, and W.Y. Man, "A Hybrid Algorithm of Differential Evolution and Machine Learning for Electromagnetic Structure Optimization," *2017 32nd Youth Academic Annual Conference of Chinese Association of Automation (YAC)*, Hefei, pp. 755–759, 2017.
14. D.R. Prado, M. Arrebola, M.R. Pino, R. Florencio, R.R. Boix, J.A. Encinar, and F. Las-Heras, "Efficient Cross Polar Optimization of Shaped-Beam Dual-Polarized Reflect Arrays Using Full-Wave Analysis for the Antenna Element Characterization," *IEEE Trans. Antennas Propag.*, vol. 65, no. 2, pp. 623–635, 2017.

15. S.K. Jain, "Bandwidth Enhancement of Patch Antennas Using Neural Network Dependent Modified Optimizer," *Int. J. Microw. Wirel. Technol.*, vol. 8, no. 7, pp. 1111–1119, 2016.

16. B.A. Sonkamble, and D.D. Doye, "Use of Support Vector Machinesthrough Linear-Polynomial (LP) Kernel for Speech Recognition," *International Conference on Advances in Mobile Network, Communicationand Its Applications*, pp. 46–49, 2019.

17. C. Perera, A. Zaslavsky, P. Christen, and D. Georgakopoulos, "Context Aware Computing for The Internet of Things: A Survey," *IEEE Commun. Surv. Tutor.*, vol. 16, no. 1, pp. 414–454, 2020.

18. J. Zhang, Z. Zhan, Y. Lin, N. Chen, Y. Gong, J. Zhong, H.S.H. Chung, Y. Li, and Y. Shi, "Evolutionary Computation Meets Machine Learning: A Survey," *IEEE Comput. Intell. Mag.*, vol. 6, no. 4, pp. 68–75, Nov. 2011.

19. C. Gianfagna, H. Yu, M. Swaminathan, R. Pulugurtha, R. Tum-Mala, and G. Antonini, "Machine-Learning Approach for Designof Nanomagnetic-Based Antennas," *J. Electron. Mater.*, vol. 46, no. 8, pp. 4963–4975, 2017.

20. P. Robustillo, J. Zapata, J.A. Encinar, and J. Rubio, "ANN Characterization of Multi-Layer Reflect array Elements for Contoured-Beam Space Antennas in the Ku-Band," *IEEE Trans. Antennas Propag.*, vol. 60, no. 7, pp. 3205–3214, 2012.

21. V. Richard, R. Loison, R. Gillard, H. Legay, and M. Romier, "Loss Analysis of a Reflect Array Cell Using ANNs with Accurate Magnitude Prediction," *2017 11th European Conference on Antennas and Propagation (EUCAP)*, pp. 2396–2399, 2017.

22. D.R. Prado, J.A. López-Fernández, M. Arrebola, and G. Goussetis, "Efficient Shaped-Beam Reflectarray Design Using Machine Learning Techniques," *2018 15th European Radar Conference (EuRAD)*, Madrid, pp. 525–528, 2018.

23. R. Florencio, R.R. Boix, and J.A. Encinar, "Enhanced MoM Analysisof the Scattering by Periodic Strip Gratings in Multilayered Substrates," *IEEE Trans. Antennas Propag.*, vol. 61, no. 10, pp. 5088–5099, 2013.

24. N.T. Tokan, and F. Gune, "Support Vector Characterization of the Microstrip Antennas Based on Measurements," *Prog. Electromagn. Res. B*, vol. 5, pp. 49–61, 2008.

25. J. Tak, A. Kantemur, Y. Sharma, and H. Xin, "A 3-D-Printed W-Band Slotted Waveguide Array Antenna Optimized Using Machine Learning," *IEEE Antennas Wirel. Propag. Lett.*, vol. 17, no. 11, pp. 2008–2012, Nov. 2018.

Chapter 7

Machine learning technique for antennas design and analysis

Rachana Kamble and Amar Nayak
Technocrats Institute of Technology Excellence, Bhopal, India

7.1 INTRODUCTION

A device which can transmit and receive electromagnetic waves is called an antenna. At the transmission end, it accepts electromagnetic waves from a transmission line and uses free space to radiate. At the reception end, it gathers electromagnetic waves from an incident wave and uses the transmission line to send. An antenna converts electromagnetic waves into electrical signals and vice versa. An antenna can be used in radio broadcasting and telecommunications to wirelessly receive and send signals. It is typically used in telecommunications and radio broadcasting to send or receive signals wirelessly. The process of optimizing and creating electrical properties and physical structure of an antenna to desired performance characteristics is referred to as antenna design [1]. The factors to take into consideration in antenna design are impedance matching, frequency range, size constraints, radiation pattern, and gain.

An array antenna or phased array antenna is arranged in a specific configuration or pattern and consists of multiple individual antenna elements. An array antenna is used in wireless networks, radar systems, radio astronomy, aerospace, satellite communication, and telecommunications [2]. To design array antenna, multiple individual antenna elements are arranged in the form of array. An antenna designed to be attached or integrated into wearable devices like clothing, accessories, smart watches, headgear, or fitness trackers is called wearable antenna. A wearable antenna provides wireless communication with comfort and portability. Designing a wearable antenna required electromagnetic simulation tools and expertise in antenna engineering [3]. A textile-integrated or fabric antenna or textile antenna is an antenna specifically designed to be integrated into wearable textiles or textile materials or smart textile applications [4]. Designing a textile antenna requires expertise in textile manufacturing processes, antenna engineering, materials science, and textile manufacturing processes.

In traditional system, the system before the era of Artificial Intelligence (AI) will only perform action according to instructions provided. They simply read the instructions, process the instructions and provide results. The characteristics of traditional computer systems are versatility, speed, diligence, accuracy, memory and reliability. The limitations of traditional

DOI: 10.1201/9781003422440-7

computer systems are lack of reasoning, common sense, feelings, or EQ (Emotional Quotient), human dependency, IQ (Intelligent Quotient), decision-making, thinking, learning capabilities, and idea presentation. Some problems required automation and experience, so that without human intervention it will perform tasks intelligently with experience [5].

The AI-based computer systems will have all the characteristics of traditional computer systems, but it will also have reasoning, common sense, thinking, learning, IQ, EQ, decision making capabilities without human intervention. If we can develop a program that teaches computers how to learn, it will solve most real-life problems [6]. The automatic recognition of spoken words, object detection and image recognition, sales prediction, and the classification of attacks in cybersecurity are among examples of self-learning and experience. Nowadays, AI technologies are developing rapidly and have produced useful intelligent systems, but the goal of achieving human intelligence is still a very distant one.

In this fifth generation i.e., the age of AI, the computers are learning from experience and can make new levels of applications which make human life better. The AI-enabled machine should have capabilities such as computer vision, automated reasoning, robotics, natural language processing, machine learning, and knowledge representation [5]. Due to AI and ML models, computers can learn like people and apply knowledge and experience to solve complex tasks easily. Many computers, such as Open AI's ChatGPT, IBM Watson, and the Google search engine, were developed using ML models. As research in ML continues to grow, it will play an important role in the development of computer science and information technology.

Among the benchmark achievements in computer science that can be attributed to ML are Google Assistant and search engines, the Facebook and Instagram facial recognition systems, Open AI ChatGPT, and many other applications. ML addresses the problem of how to create computer programs that improve their performance on certain tasks through experience. Machine learning algorithms have proven practical in most applications, such as predictive analytics in healthcare and sales, and recommendation systems in e-commerce [7].

The analysis of antennas involves studying their performance and behavior. The ML techniques for antenna analysis include Reinforcement Learning, Random Forests, Support Vector Machines, and Genetic Algorithms. Array antennas analysis techniques include Array Gain and Efficiency Analysis, Array Calibration and Error Analysis, Mutual Coupling Analysis, Array Synthesis and Optimization, Beam Forming Analysis and Radiation Pattern Analysis. Antenna modeling software and electromagnetic field solvers can be used to analyze wearable antenna [8].

The remainder of the chapter is organized as follows. Section 7.2 provides an introduction to AI, machine learning, and deep learning, applications of machine learning, types of ML, and the relationship between AI, ML, and

DL. Section 7.3 provides a literature review, with a summary of machine learning and deep learning techniques being applied to antenna design. It also provides a summary of machine learning and deep learning techniques applied to antenna analysis. Section 7.4 provides an overview of antenna design, antenna design requirements, machine learning technique for antenna design, and deep learning technique for antenna design. Section 7.5 provides an overview of antenna analysis, machine learning technique for antenna analysis, deep learning technique for antenna analysis and details of software tools for antenna analysis.

7.2 BACKGROUND

This section provides an introduction to artificial intelligence, an introduction to machine learning, an overview of deep learning, applications of machine learning in various fields, types of machine learning, and the relationship between AI, ML, and DL.

7.2.1 Introduction to Artificial Intelligence

The study of Artificial Intelligence (AI) is concerned with how to get computers to do tasks more like humans. AI provides intelligent behavior in artifacts. Intelligent behavior involves learning, action, reasoning, communicating, and perceiving in complex tasks. The goal of AI is to develop a computer system that can understand the behavior of humans, machines, or animals [7]. Tasks that are easy for humans to perform but which are difficult to describe formally offer a real challenge to AI. These include problems that we solve intuitively, such as automatically recognizing spoken words or faces in pictures. AI uses computers to mimic human behavior and intelligence using models, data, and algorithms. AI works automatically and predicts tasks normally done by humans with greater time savings, precision, reduced bias, cost, and accuracy.

The task of AI is to create intelligent entities from real-world objects that gather and understand experiential knowledge without human intervention. The biggest challenge for AI is how to incorporate the immense amount of intuitive and subjective knowledge from the real world into a computer system. AI has a wide range of applications in almost every field, including transportation, manufacturing, healthcare, entertainment, education, banking, finance, agriculture, and the military. AI can also be used to automate repetitive tasks, improve decision-making, and reduce costs [9]. AI understands existing problems and uses data to find solutions to those problems. It can be considered the science and technology of surpassing human intelligence, such as sensing, reading, experiencing, and observing, and also replicating it in computer systems.

7.2.2 Introduction to machine learning

Machine learning (ML) is a branch of computer science which works around building systems that have the ability to learn without being explicitly programmed to do so. ML teaches machines how to carry out tasks by themselves. Instead of giving instructions we give data to computers and tell them to figure out for itself. The ML technique predicts the future based on past experiences [8, 10]. AI systems acquire knowledge from raw data by extracting useful patterns and learn from these patterns. The ML techniques learn from experience to solve real-world complex tasks, make decisions and predictions that appear subjective.

A feature is a piece of information included in the representation. The AI tasks use an appropriate set of features, and then applying these features to ML algorithms to solve a task. For example, the task of speaker identification from sound requires feature such as the size of the speaker's vocal tract [11]. This feature helps to identify that the speaker is a child, woman, or man. Logistic regression, a simple ML algorithm uses feature such as the absence or presence of a uterine scar to recommend cesarean delivery. Naive Bayes, a simple ML algorithm, can classify e-mail as spam or not according to the text present in e-mail. Machine learning algorithms performance depends on feature extraction and data representation [12].

7.2.2.1 Machine learning applications

The machine learning applications are listed below. Figure 7.1 represents applications of machine learning in various fields.

The machine learning can employed to develop chatbots (Mitsuku, Replica), a self-driving car (Google), virtual and augmented reality systems (Microsoft hololens, HTC VIVE), social media services (Facebook, Instagram), service and home robots, personal assistants (Google Now, Siri, Cortana), search engines (Google, Bling), product recommendation systems (Amazon, Flipkart, Netflix, YouTube), anomaly detection systems (fraudulent transaction, behavior analysis), spam filtering (Gmail), and predictive analysis (order return prediction, sales prediction). ML can also be used in healthcare, education, transport, public safety and security systems, various learning paradigms, entertainment, low-resource communication and employment and workspace [13].

7.2.3 Introduction to deep learning

A subfield of machine learning called deep learning deals with algorithms inspired by the function and structure of the brain, called artificial neural networks, applied mathematics and statistics [9]. The origin of the deep learning concept lies in the concept of artificial neural networks (ANNs). Deep learning provides a model that learns to perform classification tasks directly from sound, text, video, or images [14].

IMAGE RECOGNITION

ONLINE FRAUD
DETECTION

SPEECH RECOGNITION

STOCK MARKET
TRADING

TRAFFIC PREDICTION

APPLICATION OF
MACHINE LEARNING

MEDICAL DIAGNOSIS

SELF DRIVING CARS

AUTOMATIC
LANGUAGE
TRANSLATION

EMAIL SPAM
&MALWARE
FILTERING

PRODUCTS
RECOMMENDATIONS

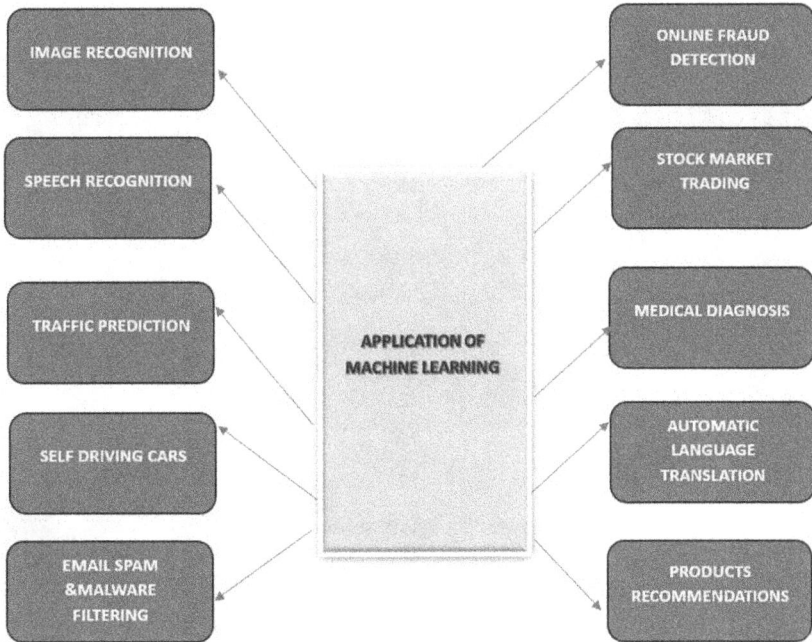

Figure 7.1 Applications of machine learning.

ANN architecture is used to implement deep learning. This ANN architecture uses nonlinear processing units arranged as a cascade of multiple
layers for transformation and feature extraction. Each successive layer uses
the output of the previous layer as input. "Deep" in deep learning refers to
the number of layers in the network—the more layers, the deeper the network. Traditional neural networks may contain only two or three layers,
while deep learning networks may have hundreds of hidden layers [15].
Deep learning approaches have also been shown to be useful for pattern
recognition, natural language processing, big data analysis, computer vision,
recommendation systems, and speech recognition. A self-driving vehicle
slows down as it approaches a crosswalk. An ATM rejects a counterfeit
banknote. A smartphone app provides an instant translation of a foreign
street sign. Deep learning is particularly well suited for identification applications such as facial recognition, text translation, speech recognition, and
advanced driver assistance systems, including lane classification and traffic
sign recognition [16].

7.2.4 The relationship between AI, ML, and DL

AI, ML, and DL are interconnected because they are related [17]. AI is a
large field that involves the development of machines that mimic human

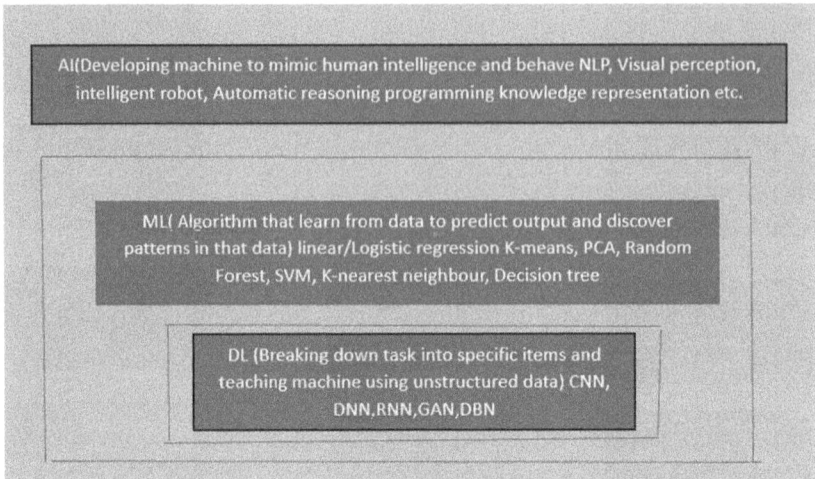

Figure 7.2 Relationships between AI, ML, and DL.

intelligence and behavior. ML is a subfield of AI that deals with automatic learning from data, and DL is a subfield of ML that involves the use of deep neural networks and unstructured data to solve complex tasks. Figure 7.2 represents the relationships between AI, ML and DL.

7.2.5 Various machine learning paradigms (algorithms, models, types)

A particular pattern by which something or someone learns is called a learning paradigm. When a data is input inside a machine, how does it learn, and how does it proceed with certain data? Figure 7.3 represents types of machine learning.

There are three types of learning paradigms associated with machine learning [18]:

(a) Supervised learning
(b) Unsupervised learning
(c) Reinforcement learning

Supervised learning: supervised learning is a machine learning method that is supervised by a supervisor. The algorithm already knows the correct answer, on the training data it iteratively makes predictions; if the predictions are not correct it is corrected by the supervisor. In supervised learning a function maps input to an output depends upon input–output pairs [19]. That is called supervised learning because the process of the algorithm learning from the training dataset can be viewed as a teacher supervising the learning process. The supervised learning algorithm contains the desired

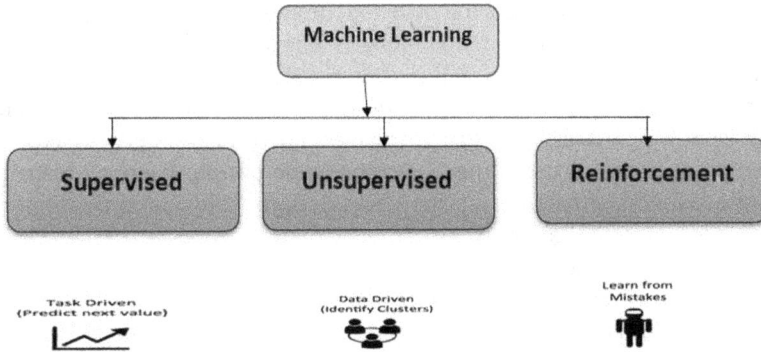

Figure 7.3 Types of machine learning.

solutions, also known as labels, in the training data. A typical supervised learning task is classification. Spam filtering and car price prediction are good examples of supervised learning [20].

Unsupervised learning: unsupervised learning algorithms process the data without a supervisor. The input to the unsupervised learning algorithm is uncategorized and unlabeled data without prior training. Semantic clustering, recommendation systems, and market basket analysis are good examples of unsupervised learning.

Reinforcement learning: reinforcement learning consists of an agent observing the environment, selecting actions, and then performing actions in return for rewards if the action is positive or a penalty if the action is negative. The algorithm then self-learns the best strategy, also referred to as a strategy, to obtain the greatest reward. Robot actions and DeepMind's AlphaGo program are good examples of reinforcement learning [21].

7.3 LITERATURE REVIEW

This section provides a literature review related to antenna design and analysis using machine learning and deep learning techniques. It also provides a summary of machine learning and deep learning techniques applied in antenna design, and a summary of machine learning and deep learning techniques applied in antenna analysis.

7.3.1 Related techniques to design antenna using machine learning

A potential future research direction in the field of antenna design is also highlighted. Article [1] discuss the use of machine learning algorithms leading to enhanced efficiency and performance and to an optimization of antenna design parameters [3]. They explore the use of genetic algorithms and reinforcement learning for antenna design optimization. Authors of [2] explore the efficient antenna designs automatically using Generative Adversarial Networks (GANs). Based on the existing antenna designs technique, GANs generate new antenna designs to improve the exploration of design space [4]. Article investigates the use of deep neural networks to solve antenna design tasks, including electromagnetic optimization problems [5]. The effectiveness of deep neural networks in optimizing antenna geometries is also demonstrated in this work.

Article [6] proposes a surrogate-assisted optimization framework using convolutional neural networks to expedite the antenna design process. The model uses results from past simulations to learn and guide future optimization. Authors of [7] explores the application of various machine learning algorithms, such as Random Forest and Support Vector Machines (SVM), to the optimization of antenna design parameters [17]. authors of [8] antenna design optimization using an evolutionary deep intelligence approach To achieve improved antenna performance the model combines deep learning techniques with evolutionary algorithms. Article [9] discuss multi-objective antenna design optimization using reinforcement learning algorithms, where multiple performance metrics are considered simultaneously. Table 7.1 shows the applications and benefits of the machine learning technique used to design antenna.

7.3.2 Related techniques to design antenna using deep learning

Article [22] proposes models to expedite the design process for electrically large antennas using a deep learning surrogate method. The authors demonstrate that reducing the need for expensive simulations and neural networks can efficiently predict the electromagnetic behavior of antennas. [23] presents approach for designing metasurface antennas based on generative adversarial networks (GANs). GANs are used to generate desired electromagnetic properties with novel metasurface structures [24] introduces how deep neural networks can be trained on a new dataset for metasurface antennas to optimize metasurface designs for specific applications present the inverse design of 3D-printed antennas using a machine learning-based approach [24]. The authors show that while considering the constraints of the 3D printing process, machine learning models can efficiently design antennas that meet given specifications.

Explores the synthesizing antenna patterns for passive radar applications using machine learning [25]. The authors show that, to achieve specific radar requirements, using machine learning techniques can optimize antenna patterns. [26] proposes a deep learning-based approach to improve the system's performance and efficiency for antenna selection in massive multi input multi output (MIMO) systems. [27] explores the use of deep learning methods to design metasurface antennas. To achieve desired antenna

Table 7.1 Machine learning techniques to design antenna with applications and benefits

ML technique	Applications and benefits
Machine Learning in Phased Array Antenna Design	Application: Utilizing machine learning techniques for phased array antenna control, beamforming, and optimization. Benefits: ML can optimize the complex interplay of antenna elements in phased arrays, leading to higher performance and better beam patterns.
Adaptive Antenna Systems with Machine Learning	Application: To adapt antenna configurations dynamically by using machine learning based on performance requirements or changing environmental conditions. Benefits: To improved performance of adaptive antennas and can optimize their characteristics in real-time.
Machine Learning in Metamaterial Antenna Design	Application: Using machine learning with unique electromagnetic properties for the optimization and design of metamaterial-based antennas. Benefits: ML can efficiently explore, identify novel antenna designs and complex metamaterial parameter spaces.
Hybrid Optimization Approaches	Application: Combining traditional optimization methods with machine learning techniques to enhance the effectiveness and efficiency of antenna design. Benefits: Hybrid approaches improve the strengths of both traditional optimization and ML to improved designs.
Machine Learning for Multi-Objective Antenna Design	Application: Applying machine learning techniques where multiple conflicting objectives need to be optimized with multi-objective antenna design problems, simultaneously. Benefits: ML models can find Pareto-optimal solutions efficiently and handle the trade-offs between different objectives.
Surrogate Modeling with Machine Learning	Application: Replace computationally expensive electromagnetic simulations using machine learning models as surrogate models. Benefits: Surrogate models accelerate the design process by providing reasonably and rapid accurate predictions.
Particle Swarm Optimization (PSO)	Application: Antenna pattern synthesis, array synthesis and parameter optimization. Benefits: PSO is capable of handling non-linear, complex, optimization problems and is computationally effective and efficient.
Antenna Parameter Estimation with Machine Learning	Application: To estimate antenna parameters such as resonant frequency or impedance from measurement data. Benefits: ML models can handle incomplete or noisy data and provide accurate parameter estimation.
Deep Learning in Antenna Design	Application: Using deep learning techniques such as CNN, RNN, neural networks or LSTM for antenna design tasks, including optimization and pattern synthesis. Benefits: Deep learning models can automate the design process and capture complex relationships in antenna design.
Genetic Algorithms (GAs)	Application: Antenna design optimization for metrics such as directivity, bandwidth, or gain to achieve desired performance. Benefits: GAs is effective in finding optimal antenna configurations and searching the vast design space for better antenna design.

properties and radiation patterns, the authors demonstrate that neural networks can optimize metasurface structures. [28] discusses how machine learning can improve characterization techniques and enhance the antenna design process. The authors present a framework for using machine learning algorithms to enhance measurements and optimize antenna design.

Introduce antenna design optimization, a deep learning-based approach with account manufacturing constraints. The proposed method can find optimal antenna configurations and efficiently search the design space [29] explores how deep learning techniques can optimize the operation and deployment of antenna system. The authors also focus on intelligent reflecting surface (IRS)-assisted wireless communication systems, in order to enhance wireless communication performance [30]. [31] presents monopole antenna design case study using deep learning. The authors demonstrate the effectiveness and feasibility of convolutional neural networks for predicting antenna properties. Table 7.2 represents a summary of machine learning and deep learning techniques applied in antenna analysis with given applications and benefits.

7.4 ANTENNA DESIGN

This section provides an overview of antenna design, and antenna design requirements. It also provides a machine learning technique for antenna design, and a deep learning technique for antenna design.

7.4.1 Antenna design overview

Antenna design requires knowledge of manufacturing processes, materials science, radio frequency (RF) engineering, and electromagnetic engineering. Antenna design also requires a balance between mechanical considerations, practical constraints to create an antenna and electrical performance to meets the requirements of the intended application. The antenna design stages are Requirements Analysis, Conceptual Design, Simulation and Modeling, Optimization, Prototyping and Testing, and Manufacturing and Production [32].

Requirements Analysis: Requirements Analysis is the first step in design of wearable, textile or array antenna. Requirements Analysis determines gain, size, bandwidth, radiation pattern, polarization, environmental constraints, and desired frequency range.

Conceptual Design: After requirement analysis the next step in antenna design is conceptual design in which, according to the requirements, antenna configurations and antenna types are considered. It depends on the type of application in which an antenna can be used. Conceptual design determines the most suitable design after evaluating antenna structures.

Table 7.2 Summary of machine learning and deep learning techniques applied
in antenna analysis

DL Technique	Applications and benefits
Genetic Algorithms (GAs)	Application: To achieve desired performance metrics (e.g., radiation pattern, gain) for antenna design optimization. Benefits: GAs is effective in finding optimal antenna configurations and searching the vast design space.
Support Vector Machines (SVMs)	Application: Based on features extracted, antenna type classification from impedance data or radiation patterns. Benefits: SVMs are well-suited for classification tasks with non-linear or separable data.
Principal Component Analysis (PCA)	Application: For antenna data analysis provides dimensionality reduction. Benefits: PCA helps in preserving important information while reducing the complexity of data.
Clustering Algorithms	Application: Based on similarity grouping antennas in radiation patterns or other characteristics. Benefits: Clustering helps in understanding antenna behavior and identifying patterns in different groups.
Convolutional Neural Networks (CNNs)	Application: Antenna pattern classification and recognition based on images radiation pattern. Benefits: CNNs can generalize well to unseen patterns and capture spatial features in antenna patterns.
Particle Swarm Optimization (PSO)	Application: Antenna array synthesis and parameter optimization. Benefits: PSO is computationally efficient and is capable of handling non-linear, complex optimization problems.
Gaussian Processes (GPs)	Application: Surrogate modeling for faster electromagnetic simulations in antenna analysis and design. Benefits: GPs can provide quantify uncertainty and probabilistic predictions in antenna parameters.
Extreme Learning Machines (ELM)	Application: Antenna optimization and pattern analysis. Benefits: ELM can achieve good generalization performance and fast training for certain tasks.
Random Forests (RF):	Application: Antenna parameter prediction and estimation, such as resonant frequency, efficiency, or gain. Benefits: RF can provide robust predictions and handle high-dimensional datasets.
Deep Reinforcement Learning (DRL)	Application: Advanced adaptive and beamforming antenna control in dynamic environments. Benefits: DRL allows antennas to learn adapt to changing conditions and control complex policies.
Reinforcement Learning (RL)	Application: Smart antenna and adaptive systems, adaptive array control and beamforming. Benefits: Enables antennas to optimize and adapt their configurations according to changing environmental conditions.

Simulation, Modeling and Optimization: After determining the conceptual design, specialized software tools are used for the simulation and modeling of antennas. Specialized simulation electromagnetic software helps analyze the antenna's performance, including impedance matching, radiation pattern, and efficiency. Simulation and modeling helps to optimize performance and refine the design of the antenna. The simulation and modeling are performed iteratively with different parameter values to improve the performance of antenna design. The antenna parameters which can be adjusted during simulation and modeling are material properties, shape, or dimensions to achieve application-specific desired characteristics such as radiation pattern, or maximum gain [33].

Prototyping and Testing: Simulation and modeling helps in finalizing the antenna design suitable for a specified application. After finalizing the design of the antenna, a physical prototype of the antenna is developed and tested. Different tests are also performed during the physical design of antenna. Measurements are taken to verify antenna performance and validate the simulation results. During testing, if the results are not as required then necessary adjustments in antenna parameters are made to get desired results [34].

Manufacturing and Production: After prototyping and testing the antenna design, if it meets the desired performance criteria, it is finalized and, using appropriate manufacturing techniques, it can be send to the manufacturing and production department for mass production. The manufacturing process may involve specialized antenna manufacturing processes, metal fabrication, or printed circuit board (PCB) fabrication.

7.4.2 Antenna design requirements

The specific requirements for designing an antenna can vary depending upon the desired performance characteristics and the intended application. However, common requirements during the antenna design process are bandwidth, frequency range, size and form factor, gain, environmental considerations, radiation pattern, power handling, polarization, cost and manufacturing feasibility, and impedance matching [32].

Bandwidth: Bandwidth is an important requirement in an antenna design process. The antenna should have adequate bandwidth to fulfill the requirement of desired frequency range. To operate within a broad frequency range or effectively across multiple frequencies, wide bandwidth is a good selection for antenna design [35].

Frequency range: The intended application must have the desired frequency range, to operate effectively while designing an antenna frequency should be in specified frequency range. The effective selection of frequency range ensures that the antenna can effectively receive and transmit signals without loss at the desired frequencies.

Size and Form Factor: The proper selection of form factor and physical size of the antenna can be important in applications with space constraints. The size of antenna should be optimized to maintain desired performance characteristics.

Gain: The ability of an antenna to concentrate or focus the received or radiated power in a particular direction is termed as gain. The gain of antenna is typically influenced by the antenna's configuration, shape, and size.

Environmental Considerations: The antenna design should consider environmental conditions such as exposure to harsh elements, humidity, or extreme temperatures in order to ensure reliability and durability. An environmental consideration improves the gain and performance and reduces the power loss. The radiation pattern may be directional (focused radiation in a specific sector or direction) or unidirectional (uniform radiation in all directions). During antenna design, depending upon the application, the antenna may require a specific radiation pattern. The radiation pattern is defined as how the antenna receives or radiates electromagnetic waves in different directions [36].

Power Handling: The antenna should be capable of handling power levels without causing damage, degradation, or distortion to the connected components or antenna.

Polarization: The orientation of the electromagnetic waves it receives or radiates is called antenna polarization. The antenna design should consider the optimized polarization requirements of the intended application to ensure efficient reception or communication of signals.

Cost and Manufacturing Feasibility: The antenna design requirements should also consider manufacturing feasibility or the cost of antenna for a specific application. The antenna design should be feasible to manufacture and economically viable within the desired cost constraints. The manufacturing feasibility and cost consideration involves minimizing complexity where possible, choosing appropriate manufacturing techniques and materials during antenna design [37].

Impedance Matching: During antenna design, proper impedance matching can also improve the performance of the antenna for specific applications. Proper impedance matching helps to optimize power transfer and minimize signal reflections. For minimizing the loss and efficient power transfer between connected the receiver/transmitter and the antenna, the impedance of the system or transmission line should be matched to the impedance of the antenna.

These antenna design requirements may be different for different applications such as broadcasting systems, radar, satellite systems, or wireless communication. The antenna design process should select the proper configuration and balance these requirements in order to manufacture antenna. The proper selection of these requirements improves the desired performance goals; and optimizes the system while taking into consideration the practical constraints.

7.4.3 Machine earning techniques for antenna design

Machine learning techniques can be used in the design of antennas to automate the design process, optimize parameters, reduce noise and improve performance. The ML techniques used in antenna design are reinforcement learning, neural networks, genetic algorithms, particle swarm optimization (PSO) [3], transfer learning, Gaussian process regression, support vector machine etc [33].

Neural Networks: Neural networks can be used to design antenna, to improve their performance and optimize antenna parameters. It can be used to build the relationship between performance metrics and antenna design parameters to optimize configuration parameters. Neural networks can assist in the proper selection of parameters, such as radiation characteristics, material properties, and antenna geometry. Neural networks can assist in this process by learning from large datasets and providing insights into the relationships between different design variables and antenna performance metrics. Neural networks can train on antenna parameter datasets during antenna design to make predictions for new designs. The neural network can predict patterns, gain, antenna geometry or impedance matching to improve the performance of antenna.

Genetic Algorithms: Genetic algorithms (GAs) are an optimization technique inspired by the process of natural evolution and selection to large search space for optimal antenna designs. GAs are used in antenna design to find near-optimal or optimal solutions in a complex design space that meet specific performance criteria. GAs used fitness functions with varying parameters to evaluate their performance and generate a population of antenna designs. GAs' optimal solution is combined and selected for antenna designs. GAs can be used to optimize antenna design parameters such as spacing, length, shape, and width in order to achieve the desired performance characteristics of antenna. GAs' success depends on encoding scheme, the proper formulation of the fitness function, and other algorithmic parameters.

Particle Swarm Optimization (PSO): Particle swarm optimization (PSO) is a population-based algorithm that can be used for optimization in antenna design to achieve optimal antenna configurations. PSO in antenna design can be used to optimize parameters like feed position, excitation coefficients to maximize gain, antenna geometry, achieve desired radiation pattern or minimize reflection coefficient. PSO is an optimization technique inspired by the collective behavior of fish schools or bird flocks [13]. PSO in antenna design can be used to improve performance metrics such as impedance matching, radiation pattern characteristics, gain, bandwidth, or other specific requirements. PSO in antenna design can be applied to various antenna types, including reflector antennas, wire antennas, array antennas and patch antennas. The basic principle of PSO involves maintaining a particles, i.e. a population of potential solutions, moving in search space for an optimal solution. Potential antenna configuration is represented by each particle's position and its movement is influenced by the best-known position of the entire swarm and its own best-known position. The positions of particles iteratively update based on these influences until a termination condition is met or the optimal solution is found. The objective function is customized according to performance requirements and the specific antenna design goals. Based on the chosen metrics, the fitness function is designed to improve the antenna's performance.

Gaussian Process Regression: Gaussian process regression, a probabilistic machine learning technique, is commonly used for regression tasks. Gaussian process regression can model the relationship between performance metrics and antenna design parameters. It is useful for modeling non-linear and complex relationships. It is based on Gaussian processes, which are non-parametric, and flexible models that can capture complex patterns in data. Gaussian process regression can be used for time series analysis and antenna design optimization, and can also be used to create surrogate models and function approximation. These models can then be utilized for sensitivity analysis or optimization.

Transfer Learning: Transfer learning is a machine learning technique that leverages knowledge gained from previously trained models and applied to another related task to accelerate the design process. Transfer learning can be applied to various aspects of antenna design, such as geometry optimization, impedance matching or radiation pattern prediction. Transfer learning allows for improved performance and faster convergence when training models for antenna design by using pre-trained models on a dataset or related task. In antenna design, transfer learning can utilize previous experience and knowledge from existing antenna designs to improve the effectiveness and efficiency of designing new antennas. The efficiency of antenna design processes can significantly enhance using transfer learning, especially in situations where the available dataset is either time-consuming to acquire or limited. By using existing experience and knowledge, transfer learning can reduce the overall design cycle time, improve the quality of designs, and expedite the design exploration.

7.4.4 Deep learning techniques for antennas design

Deep learning techniques can be used in the design of antennas to enhance or automate creating novel antenna designs and the process of optimizing antenna parameters. Commonly used deep learning techniques for antennas design are neural architecture search, generative adversarial networks, auto encoders, reinforcement learning, and hyperparameter optimization [34].

Generative Adversarial Networks (GANs): Generative adversarial networks (GANs) are deep learning models that can be used to generate novel antenna designs. GANs consist of two components, a discriminator and a generator which are trained in an adversarial setting. GANs can be used to optimize existing designs or to generate novel antenna designs with specific characteristics by training a generator network to create synthetic antenna patterns. The discriminator network evaluates the similarity and quality to real antenna designs while the generator learns to produce new antenna designs. GANs can help generate and explore antenna designs with desired performance characteristics. GANs can automatically speed up the design process, optimizing performance metrics and generating innovative designs. The quality of the generated antenna designs relies on the diversity and quality of the training dataset, as well as the training parameters and architecture of the GAN model.

Reinforcement learning (RL): Reinforcement learning (RL) can help optimize and automate the process of designing an antenna by iteratively improving and learning antenna configurations based on rewards and feedback. RL algorithms autonomously optimize antenna designs by training the agent. The agent interacts with the real or simulated environment and provides feedback on the performance of the process of antenna design. The agent learns to select antenna parameters through a trial-and-error process and get the desired performance characteristics. RL steps in antenna designs include the following: define the problem, formulate the RL problem, state space, action space, reward function, learning algorithm, training the RL agent, testing and evaluation and iterative refinement.

Autoencoders: Autoencoders are neural network models mainly used for unsupervised learning tasks, such as data compression, dimensionality reduction and feature extraction. In antenna design, autoencoders can be used for feature extraction, generate novel antenna designs; perform design optimization, and dimensionality reduction. The model creates a compressed representation of the designs and learns to extract essential features by training an autoencoder on a dataset of antenna designs. This helps in generating new designs with similar features, identifying important design parameters and exploring the design space. The success of autoencoders in antenna design depends on the diversity and quality of the training dataset, as well as the training parameters and architecture of the autoencoder model [38].

Neural Architecture Search (NAS): Neural Architecture Search (NAS) is a deep learning model used to automatically search for optimal neural network architectures. NAS can be used in antenna design to optimize antenna performance, discover novel and efficient designs, explore a wider design space, and automate the process antenna parameter configurations or the automatic discovery of optimal antenna architectures. It allows for more effective and efficient antenna design by automating the process of high-performing designs searching. NAS can automatically discover efficient and effective antenna configurations by using evolutionary algorithms or reinforcement learning and defining a search space of possible antenna designs.

Surrogate Modeling: Surrogate modeling, also known as response surface modeling, or metamodeling, is a deep learning technique used to approximate the behavior of a complex system using a simplified model. The accuracy of the surrogate model depends on the chosen modeling technique and the representativeness, diversity, and quality of the training dataset. Regular validations and updates of the surrogate model are recommended to ensure its effectiveness, reliability and efficiency throughout the design process. Deep learning models, for example neural networks, can serve as surrogate models to approximate the behavior of antennas. By training on a dataset of antenna measurements or simulations, for various parameter configurations the surrogate model can make fast predictions of antenna performance. Surrogate models can enable efficient exploration of the design space and accelerate the optimization process. In antenna design, surrogate modeling can optimize antenna performance, explore the design space and expedite the design process. The advantages of surrogate modeling in antenna design are enabling efficient exploration of the design space, facilitating optimization processes and reducing the computational burden associated with expensive simulations.

Hyperparameter Optimization: Hyperparameters are parameters that are set by the machine learning engineer prior to training but rather not learned during the training process. Hyperparameter affect the performance and behavior of the deep learning algorithm, and proper tuning them is essential for obtaining the optimized results. Hyperparameters can be optimized using deep learning techniques in antenna design. The optimal values of hyperparameters, such as regularization terms, activation functions, or learning rates, can be identified in antenna design to improve the performance of deep learning models by using techniques such as Bayesian optimization or genetic algorithms. Hyperparameter helps in fine-tuning the deep learning algorithms or achieving optimal performance in antenna design tasks.

The deep learning techniques for antenna design require careful consideration of the validation technique, available data, and the need for domain expertise.

7.5 ANTENNA ANALYSIS

This section provides an overview of antenna analysis, the machine learning technique for antenna analysis, the deep learning technique for antenna analysis and details of software tools for antenna analysis.

7.5.1 Overview of antenna analysis

Antenna analysis is the process of understanding and evaluating the performance, characteristics, and behavior of antennas. Antenna analysis is typically performed using electromagnetic field solvers, numerical methods, and computer simulation software. Physical measurements are also involved in antenna analysis using antenna test ranges or anechoic chambers. Antenna analysis involves examining and studying various aspects of antennas to ensure they adhere to desired specifications, perform optimally, and meet specific design objectives [35]. The analysis of antenna includes efficiency analysis, radiation pattern analysis, polarization analysis, impedance analysis, 3D pattern visualization, Gain analysis, and bandwidth analysis.

Efficiency Analysis: Efficiency analysis represents how well the antenna converts electrical power into radiated electromagnetic waves. High efficiency is important to improve overall performance and reduce energy losses.

Radiation Pattern Analysis: The directional properties of an antenna are described by radiation patterns. Radiation patterns show how the antenna receives or radiates electromagnetic energy in different directions. Analyzing the radiation pattern helps to determine the antenna's side lobes, beamwidth, gain, directivity, nulls and coverage area.

Impedance Analysis: Impedance analysis represents how well the antenna matches the impedance of the system, or of the transmission line to which it is connected. A good impedance match is crucial to minimize signal reflections and losses and maximize power transfer.

3D Pattern Visualization: 3D pattern visualization represents a graphical representation of the antenna's radiation pattern in three-dimensional space. This visualization helps engineers understand the antenna's coverage area and its assessing spatial characteristics.

Gain Analysis: Gain analysis represents the ability of the antenna to concentrate radiated energy compared to an isotropic radiator in a specific direction. High gain is desirable for directional antennas to enhance signal strength and communication range.

Polarization Analysis: Polarization analysis analyzes the polarization characteristics of the antenna, like elliptical polarization, circular, or linear. Proper polarization matching is essential for optimal signal transmission and reception.

Bandwidth Analysis: The analysis of range of frequencies over which the antenna operates effectively is termed as bandwidth analysis. To support multiple frequency channels for communication systems wide bandwidth is desirable of antenna.

Antenna analysis is a critical step in the antenna engineering process and design to ensure the reliable and efficient operation of antennas in various applications, including satellite communications, radar systems, wireless communications, and more. The observations from antenna analysis help antenna design engineers to ensure that the antennas meet specific requirements for their intended applications, troubleshoot performance issues, and optimize antenna designs.

7.5.2 Machine learning techniques for antennas analysis

Machine learning techniques can be used in the analysis of data associated and antenna performance with antennas. Machine learning techniques commonly used for antenna analysis are signal processing, quality control, pattern recognition, optimization, predictive maintenance, data-driven modeling, and anomaly detection [36].

Pattern Recognition: Machine learning algorithms can be used to classify and analyze antenna radiation patterns based on specific characteristics or features. The ML algorithm can learn to classify and identify different types of patterns like sectorial patterns, directional, or omnidirectional by training a model on a dataset of known radiation patterns.

Anomaly Detection: ML algorithms can help identify deviations or anomalies in antenna performance metrics. By training on a dataset of expected or normal performance, the algorithm can learn to recognize patterns that malfunction in the antenna system, indicating potential issues or deviating from the norm.

Signal Processing: Machine learning techniques can be applied to process and analyze signals transmitted or received by antennas. ML algorithms can be used to analyze received signals for tasks like interference detection, signal classification or modulation recognition.

Predictive Maintenance: ML algorithms can be used to predict when servicing or maintenance may be required and to analyze antenna performance data over time. By training on historical data of maintenance records and antenna performance, the algorithm can learn patterns that indicate failure or degradation and provide proactive maintenance recommendations.

Optimization: By exploring large design spaces, ML algorithms can be used to optimize antenna performance. By using algorithms like reinforcement learning or genetic algorithms, optimal antenna parameters can be identified to achieve specific radiation pattern characteristics, minimum reflection coefficient, desired performance metrics, maximum or gain.

Data-driven Modeling: ML techniques, such as neural networks and regression models, can be used to create data-driven models of antenna behavior. These models can predict antenna performance metrics based on accurate analysis, enabling quick input parameters without the need for extensive measurements or simulations.

Quality Control: ML algorithms can also be used in quality control processes for antenna manufacturing. ML models can identify deviations, variations, or defects, from desired specifications, allowing for early corrective and detection actions by analyzing data from antenna production lines.

These ML techniques can assist in improving overall antenna system analysis, optimizing parameters, predicting maintenance needs, analyzing antenna performance, and identifying patterns.

7.5.3 Deep learning techniques for antennas analysis

Deep learning models, a subset of machine learning, can be employed in antenna analysis tasks to make predictions, classify patterns, or extract insights. Deep learning techniques commonly used for antennas analysis include convolutional neural networks, recurrent neural networks, generative adversarial networks, transfer learning, and attention mechanisms [37].

Generative Adversarial Networks (GANs): Generative adversarial networks (GANs) can be used for antenna-related tasks such as optimizing antenna designs, or generating synthetic radiation patterns or antenna analysis. The discriminator network learns to distinguish between generated and real samples while the generator network learns to generate realistic antenna designs, or patterns resulting in realistic synthetic data generation or improved designs.

Convolutional Neural Networks (CNNs): Convolutional neural networks (CNNs) can be applied to antenna-related tasks such as analyzing antenna images or radiation patterns, although it is widely used for image analysis. CNNs can automatically learn spatial and features relationships within antenna images, enabling tasks like identification of anomalies, pattern recognition, or classification of antenna types.

Recurrent Neural Networks (RNNs): Recurrent neural networks (RNNs) are suitable for tasks involving antenna performance or signals measurements over time as it is commonly used for analyzing time series data. RNNs can learn patterns in sequential data and capture temporal dependencies, enabling tasks like modulation recognition, signal classification, or predicting future antenna performance based on historical data.

Autoencoders: Autoencoders are unsupervised learning-based neural networks models that can learn by themselves and efficiently represent antenna data. They can be used for feature extraction or dimensionality reduction in antenna analysis. By training an autoencoder on antenna performance data, it can learn efficient and compressed representations of the antenna data, allowing for efficient visualization and analysis.

Transfer Learning: Transfer learning uses pre-trained deep learning models and applies them to antenna analysis tasks with limited data. Transfer learning can work efficiently on large-scale datasets. By using pre-trained fine-tuned models on antenna-specific data, deep learning models can learn relevant patterns and features, even with limited training samples. This approach can improve performance and speed up the learning process in antenna analysis tasks.

Attention Mechanisms: Deep learning models commonly used attention mechanisms to focus on important parts of the input data. Attention mechanisms in antenna analysis can be used to highlight specific features or regions in antenna signal or images data that contribute most to the desired task, such as classification or anomaly detection in antenna analysis.

These deep learning techniques offer powerful capabilities for optimization, anomaly detection, prediction, antenna analysis, automated classification, and enabling the extraction of valuable insights. However, it's important to have a careful validation to ensure accurate and reliable outcomes and sufficient amount of labeled.

7.5.4 Antenna design and analysis software tools

Widely used antenna simulation, design and analysis by researchers and engineers are ANSYS HFSS, CST Studio Suite, FEKO, WIPL-D, ADS, 4nec2, EZNEC, Antenna Magus, GRASP, MMANA-GAL, and MATLAB Antenna Toolbox [38].

CST Studio Suite: CST Studio Suite is widely used electromagnetic simulation software that provides capabilities for antenna optimization, analysis, and design. It offers various solver options and 3D modeling capabilities.

WIPL-D: WIPL-D is a full-wave 3D electromagnetic solver used for the optimization, analysis, and design of complex microwave and antennas structures.

ANSYS HFSS (High-Frequency Structure Simulator): HFSS is a powerful electromagnetic simulation software tool used for analyzing and designing high-frequency antennas. It offers a wide range of capabilities for various antenna structures and types.

4nec2: 4nec2 is an open-source software tool that uses the Numerical Electromagnetics Code (NEC) method to design, optimize and analyze antennas.

FEKO: FEKO is an electromagnetic simulation software tool that includes various packages for antenna design, optimization, and analysis using FEM (Finite Element Method), FDTD (Finite Difference Time Domain), and MoM (Method of Moments).

MATLAB Antenna Toolbox: MATLAB's Antenna Toolbox provides a set of tools and functions for analyzing and designing antennas using MATLAB's programming environment.

EZNEC: EZNEC is popular antenna analysis, design and modeling software based on the Numerical Electromagnetics Code (NEC) method. It provides a graphical interface for analyzing and designing antennas.

ADS (Advanced Design System): ADS is electronic design automation software tools widely used for designing RF/microwave circuits and antennas.

GRASP: GRASP (General Reflector Antenna Software Package) is a widely used specialized software tool for designing reflector antennas for satellite communications.

MMANA-GAL: MMANA-GAL (MMana-Gal—An Amateur Radio Antenna Analyzer) is another open-source and widely used antenna optimization, design, and analysis tool based on the Method of Moments (MoM).

Antenna Magus: Antenna Magus is a commercial antenna design tool that consists of a database of the antennas element and allows users to export, design, and explore antenna models to other simulation software.

7.6 CONCLUSIONS

The antenna design stages are requirements analysis, conceptual design, simulation and modeling, optimization, prototyping and testing, and manufacturing and production. Antenna analysis involves examining and studying various aspects of antennas to ensure they adhere to desired specifications, perform optimally, and meet specific design objectives. The analysis of antenna includes efficiency analysis, radiation pattern analysis, polarization analysis, impedance analysis, 3D pattern visualization, gain analysis, and bandwidth analysis. This chapter provides a description of machine learning techniques employed in antenna design and analysis. The machine learning techniques used in antenna design are reinforcement learning, neural networks, genetic algorithms, particle swarm optimization, transfer learning, Gaussian process regression, support vector machine etc. Commonly used deep learning techniques for antennas design are neural architecture search, generative adversarial networks, autoencoders, reinforcement learning, and hyperparameter optimization. Machine learning techniques commonly used for antenna analysis are signal processing, quality control, pattern recognition, optimization, predictive maintenance, data-driven modeling, and anomaly detection. Deep learning techniques commonly used for antennas analysis are convolutional neural networks, recurrent neural networks, generative adversarial networks, transfer learning, and attention mechanisms.

REFERENCES

1. R. Rashmitha, N. Niran, Abhinandan Ajit Jugale, and Mohammed Riyaz Ahmed, "Microstrip Patch Antenna Design for Fixed Mobile and Satellite 5G Communications," *Procedia Comp. Sci.*, vol. 171, pp. 2073–2079, 2020.
2. Ilhem Gharbi, Rim Barrak, Mourad Menif, and Hedi Ragad, "Design of Patch Array Antennas for Future 5G Applications," In *18th International Conference on Sciences and Techniques of Automatic Control and Computer Engineering (STA)*, pp. 314–327, IEEE, 2017.
3. N. Singh, T. Chakrabarti, P. Chakrabarti, M. Margala, A. Gupta, S. B. Krishnan and B. Unhelkar, "A New PSO Technique Used for the Optimization of Multiobjective Economic Emission Dispatch," *Electronics*, vol. 12, pp. 1–13, 2023.
4. Abhishek Agarwal, Riki Patel, Arpan Desai, Trushit Upadhyaya, and Shubham Patel, "Analysis of Flexible Textile Antenna Design and Their Application for Wireless Communication: A Survey," In *Third International conference on I-SMAC (IoT in Social, Mobile, Analytics and Cloud) (I-SMAC)*, pp. 481–490, IEEE 2019.
5. Amar Nayak, and Rachana Kamble, *Artificial Intelligence and Machine Learning Techniques in Power System Automation*, Auerbach Publications, pp. 216–230, 2023.
6. Cheon-Bong Moon, Jin-Woo Jeong, Kyu-Hyun Nam, Zhou Xu, and Jun-Seok Park, "Design and Analysis of a Thinned Phased Array Antenna for 5G

Wireless Applications," *Int. J. Antennas Propag.*, pp. 652–663, 2021. https://doi.org/10.1155/2021/3039183

7. Jiaying Liu, Xiangjie Kong, Feng Xia, Xiaomei Bai, Lei Wang, Qing Qing, and Ivan Lee, "Artificial Intelligence in the 21st Century," *IEEE Access*, vol. 6, pp. 551–563, 2018.

8. Sheena Angra, and Sachin Ahuja, "Machine Learning and Its Applications: A Review," *International Conference on Big Data Analytics and Computational Intelligence (ICBDAC)*, pp. 267–775, IEEE 2017.

9. Ajay Shrestha, and Ausif Mahmood, "Review of Deep Learning Algorithms and Architectures," *IEEE Access*, vol. 7, pp. 411–421.

10. Pariwat Ongsulee, "Artificial Intelligence, Machine Learning and Deep Learning," *15th International Conference on ICT and Knowledge Engineering (ICT & KE)*, pp. 618–626, IEEE 2017.

11. Iqbal H. Sarker, *Machine Learning: Algorithms, Real-World Applications and Research Directions*, Springer, pp. 211–223, 2021.

12. He, and Li, "Deep Learning for Antenna Design: An Overview, Recent Advances, and Future Directions," *IEEE* 2020, pp. 166–175.

13. N. Singh, T. Chakrabarti, P. Chakrabarti, V. Panchenko, D. Budnikov, I. Yudaev and V. Bolshev, "Analysis of Heuristic Optimization Technique Solutions for Combined Heat-Power Economic Load Dispatch," *Appl. Sci.* vol. 13, pp. 1–19, 2023. https://doi.org/10.3390/app131810380

14. O. Koasteen, J. V. Mohanan, A. Gupta, C. Christodoulou, "Antenna Design Using a GAN-Based Synthetic Data Generation Approach", vol. 3, pp. 488–494, 2022.

15. G. Rohini, C. Gnana Kousalya, Nagendra Singh, Visal Ratnsing Patil. "Autonomous Forecasting of Traffic in Cellular Networks Based on Long-Short Term Memory Recurrent Neural Network," *Cybern. Syst. Int. J*, vol. 54, pp. 132–145, 2023. https://doi.org/10.1080/01969722.2023.2166245

16. Yi Zhao, Chaoli Sun, Jianchao Zeng, Ying Tan, Guochen Zhang, "A surrogate-ensemble assisted expensive many-objective optimization", *Knowledge-Based Syst.*, vol. 211, 2021, https://doi.org/10.1016/j.knosys.2020.106520

17. Guha, and Pandey, "Antenna Design Optimization Using Machine Learning Algorithms," pp. 511–521, IEEE 2022.

18. Y. Yin, "Evolutionary Deep Intelligence for Antenna Design," pp. 816–825, IEEE 2021.

19. Singh, and Deb, "Multi-Objective Antenna Design Using Reinforcement Learning," pp. 739–747, IEEE 2019.

20. M. N. Maud, S. M. Khallaf, and S. A. R. Abu-Bakar, "Efficient Antenna Design with Deep Learning Surrogate Models for Electrically Large Problems," *IEEE Antennas Wirel. Propag. Lett.*, pp. 510–521, 2021.

21. R. Jiang, X. Wang, S. Cao, J. Zhao, and X. Li, "Deep Neural Networks for Channel Estimation in Underwater Acoustic OFDM Systems," *IEEE Access*, vol. 7, pp. 23579–23594, 2019.

22. C. R. Vallecilla, A. S. Omar, and A. V. Diebold, "Metasurface Design Using Deep Neural Networks Trained on a Novel Dataset," *Antennas and Propagation Society International Symposium*, pp. 87–98, IEEE 2021.

23. M. H. Mirsalehi, and K. Sarabandi, "Machine Learning-Based Inverse Design for Antennas Employing 3D Printing," *IEEE Trans. Antennas Propag.*, vol. 70, pp. 210–223, 2020.

24. M. J. Luong, F. B. Trang, and H. T. Ewe, "Antenna Pattern Synthesis Using Machine Learning for Passive Radar Applications," *PIERS*, pp. 419–421, 2020.
25. H. Choubey, V. Arya, J. Singh, N. Choudhary, A. Sharma, N. Singh, "Efficient Model based on Deep Learning for the Classification of Dementia," *Int. J Recent Innovation Trends Comp. Commun.*, vol. 11, no. 10, pp. 1056–1061, 2023.
26. S. Sun, A. V. Kildishev, and V. M. Shalaev, "Deep Learning-Based Inverse Design of Metasurface Antennas," *Phys. Rev. App.*, pp. 934–941, 2017.
27. A. O. Boryssenko, R. H. M. Hafez, and A. C. Polycarpou, "Machine Learning-Enhanced Antenna Design and Characterization," *IEEE Antennas Propag Mag.*, pp. 1078–1087, 2020.
28. N. Singh and Y. Kumar, "Blockchain for 5G Healthcare Applications: Security and privacy solutions," *IET Digital Library*, vol. 1, pp. 1–30, 2021. DOI:10.1049/PBHE035E
29. Q. Wu, Y. Zeng, and R. Zhang, "Intelligent Reflecting Surface-Assisted Wireless Communications: A Deep Learning Perspective," *IEEE Wirel. Commun.*, vol. 22, pp. 391–401, 2019.
30. A. A. Alabwah, A. J. Salim, and A. M. Abdelgawad, "Deep Learning for Computational Electromagnetics: A Case Study on Monopole Antenna Design," *IEEE Access*, 2018, pp. 2987–3997.
31. Aktül Kavas, and A. Eyüp Kırık, "Antenna Design for Smart Devices," *24th Signal Processing and Communication Application Conference (SIU)*, pp. 378–385, IEEE 2016.
32. Giang Bach Hoang, Giap Nguyen Van, Linh Ta Phuong, Tuan Anh Vu, and Duong Bach Gia, "Research, Design and Fabrication of 2.45 GHz Microstrip Patch Antenna Arrays for Close-Range Wireless Power Transmission Systems," *International Conference on Advanced Technologies for Communications (ATC)*, pp. 498–509, IEEE 2016.
33. Hilal M. El Misilmani, and Tarek Naous, "Machine Learning in Antenna Design: An Overview on Machine Learning Concept and Algorithms," *International Conference on High Performance Computing & Simulation (HPCS)*, pp. 684–693, IEEE 2019.
34. A. M. Montaser, "Deep Learning Based Antenna Design and Beam-Steering Capabilities for Millimeter-Wave Applications," pp. 398–409, IEEE 2021.
35. N. Singh, and G. Kumar Prajapati, "Wireless Sensor Network Security Problems-A Review", *Int. J Emerging Technol. Adv. Eng.*, vol. 8, no. 1, pp. 77–80, 2018.
36. Christoph Maeurer, Peter Futter, and Gopinath Gampala, "Antenna Design Exploration and Optimization using Machine Learning," *14th European Conference on Antennas and Propagation (EuCAP)*, pp. 751–758, IEEE 2020.
37. Zhao Zhou, Zhaohui Wei, Jian Ren, Yingzeng Yin, and Gert Frølund Pedersen, "Two-Order Deep Learning for Generalized Synthesis of Radiation Patterns for Antenna Arrays," *IEEE Trans. Artif. Intell.*, vol. 4, pp. 1–9, 2022.
38. Arjun Gupta, John Argyres, Ralph Lyndon Gesner, and Delaney Rose Heileman, "DIY Antenna Studio: A Cost-Effective Tool for Antenna Analysis," *IEEE Antennas Propag.*, vol. 63, no. 2, pp. 83–88, 2021.

Chapter 8

Design of implantable antenna for biomedical applications

Dipti Shukla

O.P Jindal University, Raigarh, India

8.1 INTRODUCTION

The communication system provides an information exchange between two sites. During communications, information is both transmitted and received. The three main components of communication are the information transmitter, the channel or medium of communication, and the information receiver. The communication system is divided into two categories based on communication channels.

Wired (Line communication)   **Wireless** (Spacecommunication)

☐ Parallel wire communication   ☐ Ground wave communication

☐ Twisted wire communication   ☐ Skywave communication

☐ Coaxial cable communication   ☐ Space wave communication

8.2 ANTENNA

An antenna is regarded as the most important component of a wireless communication system. This is a gadget that converts an electromagnetic wave to a Radio Frequency (RF) signal in order to send it into empty space. As is well known, antennas are necessary for both transmission and reception,. As a result, antennas for broadcasting and receiving radio waves are required. Antennas are thought of as the fundamental building blocks of a wireless communication system because a signal in empty space spreads as an electromagnetic wave. It therefore needs some sending and receiving endpoints. An antenna is a metallic device used to transmit or receive radio electromagnetic waves, which are offered in a variety of forms and sizes. On the roofs of domestic buildings, tiny antennas are used to view television, while larger ones are used to collect signals from satellites. The main type of antenna used by SCaN (Space Communications and Navigation) is a parabolic antenna, which has a bowl-like form and concentrates signals at one end, as shown in Figure 8.1.

DOI: 10.1201/9781003422440-8

Figure 8.1 Antenna (wireless communication system).

An antenna enables electromagnetic waves (EMW) to travel from one end to the other without the use of a wired system. At the transmitter, it transforms an RF signal into an EM wave, and at the receiver, the reverse happens, with the EMW converted into an electrical signal. Antennas do this by creating an electromagnetic wave by combining an electric and magnetic field created by the applied signal. The reason for this is because the magnetic and electric fields are perpendicular to one another. Thus

$$E.H = 0 \tag{8.1}$$

We frequently encounter the usage of wireless communication in our daily lives. Therefore, without the need for a wired system, an antenna enables the transmission of EMW from one end to the other. It performs the function of a transducer, transforming a transmitter's RF signal into an electromagnetic wave and a receiver's electromagnetic (EM) wave back into an electrical signal. Wireless communication underlies everything, from signals reaching the television for viewing a variety of programs to chatting with anyone on a mobile network. Antennas are thus an essential aspect of modern life.

8.2.1 Why is an antenna array necessary?

We are aware that a tiny linear antenna has an irregular radiating field in the direction perpendicular to its axis. This non-uniform radiation pattern is appropriate for broadcasting-related applications. Dipole antennas typically have an omnidirectional radiation pattern, whereas highly directed radiation is provided by antennas like horns, helical, slots, etc. Thus, these suites enable point-to-point communication by directing focused energy in a single direction. We are aware that increased directivity is required as the transmitter and receiver are increasingly apart from one another. This is the case since path loss also rises as the distance increases. Therefore, if attenuation is also strong across a great distance, it is probable that the signal will not reach the other end with the necessary power [1–3].

This is the rationale behind the necessity of high-gain antennas. However, there are situations when a single antenna's gain is insufficient to successfully transmit a signal by overcoming attenuation. An antenna array is utilized for this purpose.

8.2.2 Array antenna

For improved output transmission, an antenna is a specialized device that can emit certain energy in a specific direction. Antenna arrays are created by adding additional antenna elements in order to provide a more effective output. An antenna array is utilized because a single antenna has strong directivity but falls short in signal transmission to the receiver due to losses. Because of this, we require antennas with exceptionally high directional qualities in a variety of applications, which may be better by expanding the electrical dimension of the antenna. Forming antenna array elements increases the antenna's dimension without increasing the dimensions of the individual components. An overview of antenna array types and how they function in various applications is covered in this chapter.

In order to produce radiation patterns that cannot be produced by individual antennas, a set of antennas is arranged to form an antenna array. In order to broadcast or receive radio signals, a group of antennas will cooperate. Because each antenna is smaller, it is affordable to develop and maintain this antenna.

When designing the antenna array, sufficient spacing and phase must be provided. Antennas that transmit signals over very long distances must have a high directional gain because the signal deforms and distorts as it travels from one end to the other. Even if a single antenna broadcasts with high directivity, it cannot provide a signal without any losses from transmitter to receiver. Thus, the antenna array is mostly used for this purpose.

An antenna array, also known as an array antenna, is a collection of interconnected antennas used for radio wave transmission or reception [4].

A set of two or more antennas is referred to as an "antenna array," sometimes known as a "phased array." The signals from the antennas are combined or processed in order to outperform the performance of a single antenna. The antenna array can be used for the following reasons:

- To boost total profit.
- To welcome diversity.
- To eliminate interference coming from a specific direction.
- To "steer" the array to focus its sensitivity in a certain area.
- To determine the direction in which incoming signals will travel to improve the Signal Interference plus Noise Ratio (SINR).

Antennas

Antenna array
axis

d   d

Direction of Radiation

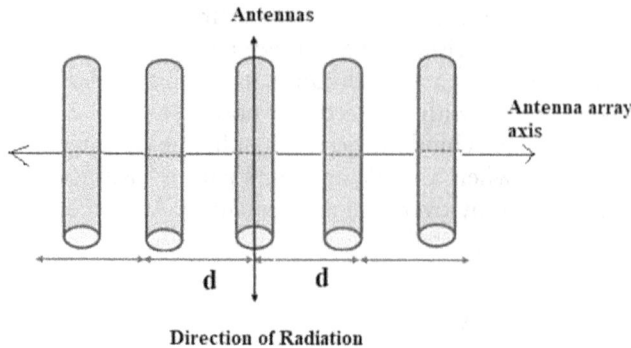

Figure 8.2 Antenna array.

8.2.3 Design of array antennas

In order to create an antenna array with a high directional gain, many antennas are arranged into a single system. In order for each antenna's independent contribution to be added up in one direction while being canceled out in all other directions, antennas within the array must be positioned appropriately and in the correct phase. The directivity of the system is enhanced by this form of set-up. A linear antenna array is a system in which all of the antennas are aligned in a straight line, as shown in Figure 8.2.

8.2.4 Operation of array antenna

An antenna array is a collection of different antenna components. In general, a half-wave dipole antenna is used in multi-element arrays. The waves are emitted above a wide angle because of the omnidirectional radiation pattern of this antenna. These antennas are simply formed in an array configuration with the correct spacing to improve their ability to emit, particularly in one direction. In order to concurrently activate these arrays, current must be supplied in the correct phase. The currents inside the various antenna elements in an array of antennas are often said to be in phase if they reach their peak value while simultaneously flowing in a comparable direction. Consequently, the spherical waves are overlaid due to interference and produce a radio wave once the antenna components are supplied with the correct phase from each member of the array. The interference in this system, which entirely depends on the waves sent by the components, can be either constructive or destructive. As a consequence, if the waves that the antenna elements emit are in phase, they may be joined together to enhance the power that is radiated [5]. The waves, however, are combined destructively to cancel each other out if the waves released by the various constituents are not in phase. Accordingly, this may result in a decrease in the radiated power.

In this way, the radiations that are released from the array's components are in phase and combine to form a directed beam with the greatest power and the ability to travel over very long distances. As a result, an antenna array's radiation pattern features a primary lobe that designates a powerful beam in a particular direction. The primary lobe of the array becomes narrower as the number of elements increases, while smaller side lobes indicate an increase in the gain offered by the antenna.

(The main lobe grows narrower and the side lobes get smaller as the number of antenna components increases, suggesting an increase in the gain the antenna offers.)

8.2.5 Kinds of array antenna

Figure 8.3 represents the types of array antenna.

8.2.5.1 Broadside array

Below is a diagram of a broadside antenna array, where similar components are parallel-arranged along the antenna axis line. Every element in this sort of arrangement is fed by a current that has a comparable phase and magnitude and is horizontally placed at an equal distance from one another. The broadside, or the direction normal to the array axis, will produce the most radiation whenever the elements in this arrangement are energized, although the other directions will also emit some radiation. As a result of radiating in both directions along the broadside, it offers a bidirectional radiation pattern. Therefore, the array axis and the plane of element location in this arrangement share the principle of radiation direction. The broadside antenna array's radiation pattern is seen in Figure 8.4.

8.2.5.2 End-fire array

The components of an end-fire antenna array layout are the same as those in a broadside set-up, but the method of excitation differs significantly between

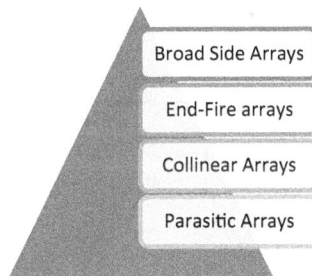

Figure 8.3 Kinds of array antenna.

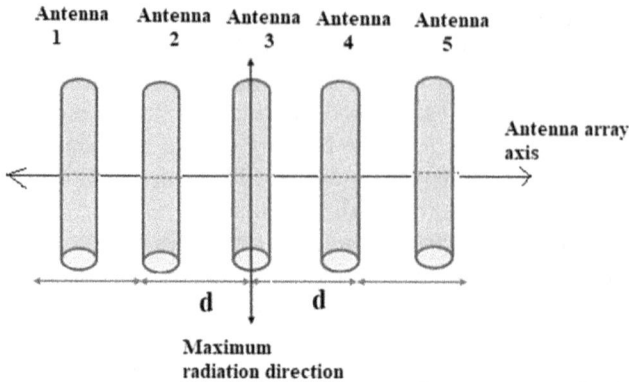

Figure 8.4 Broadside array.

the two configurations. In contrast to a broadside arrangement, where each element is fed with a current of a comparable phase, the elements in this configuration are typically fed 180 degrees out of phase. In this configuration, the array axis is where the highest radiation is obtained. Thus, this entire arrangement of similar parts is simply energized with comparable amplitude current in order to provide a unidirectional radiation pattern, but the phase changes continually down the line. Therefore, it may be said that an end-fire array produces a unidirectional radiation pattern since the antenna array's axis experiences the maximum radiation. The major separation between the parts in this arrangement in Figure 8.5 is typically a radio antenna with high directional characteristics made up of many line-arranged antenna components. Because they are suitable for high-, medium-, and low-frequency ranges, these arrays are most often employed in point-to-point communication [6].

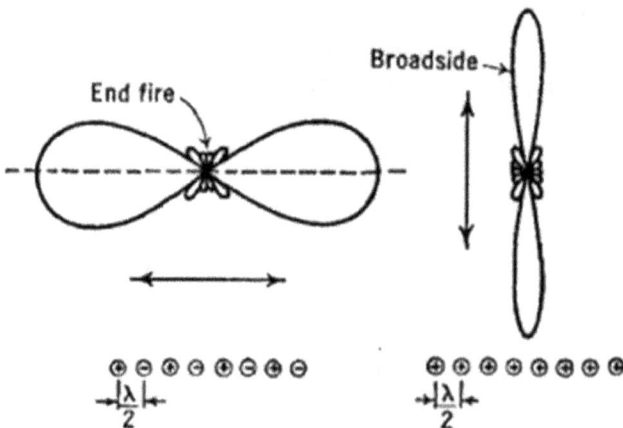

Figure 8.5 End-fire array.

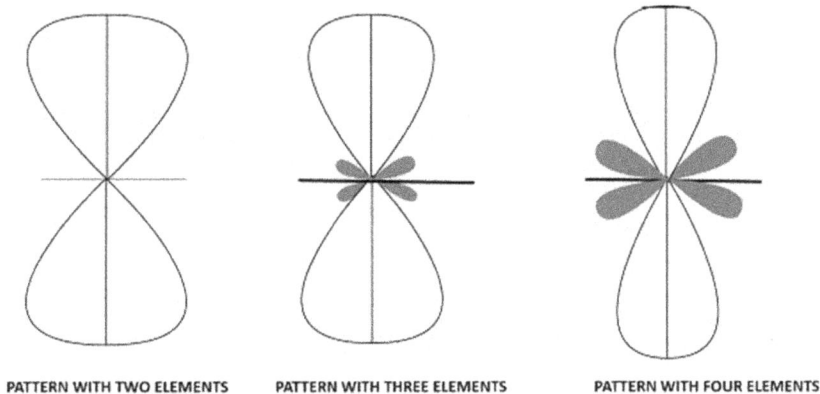

PATTERN WITH TWO ELEMENTS   PATTERN WITH THREE ELEMENTS   PATTERN WITH FOUR ELEMENTS

Figure 8.6 Collinear array.

8.2.5.3 Collinear array

Antenna elements are simply placed in a single line from one end to the other, or one after the other, in a collinear array. Therefore, the orientation of this arrangement might be either horizontal or vertical. Figure 8.6 is a diagram of a collinear array with a horizontal configuration. Currents of the same phase and magnitude are used to deliver stimulation to every antenna element. This also radiates in the normal direction to the antenna array axis, much like a broadside array. As a result, there is a connection between the collinear array's radiation pattern and the broadside antenna array. This configuration only provides the best gain when the components are separated by 0.3 to 0.5, but this might result in construction and feeding problems with the antenna array. As a result, the components are placed closer together [7].

8.2.5.4 Parasitic array

In order to get maximum directional gain without feeding every member of the array, multi-element arrays are stacked parasitically, much like parasitic antenna elements. By avoiding giving direct stimulation to each antenna array piece, this type of design just assists in addressing the feed line issue. The parasitic antenna configuration is seen in Figure 8.7. The parasitic elements, also known as non-fed elements, get their energy from the radiation that the driven element emits while it is in close proximity. As a result of the driving element's proximity to the parasitic components, electromagnetic coupling is triggered.

The parasitic elements of the antenna array rely on the excitation directed toward the driving element rather than being stimulated directly. The distance between these two elements and their tuning, therefore, defines the induced current within the parasitic element created by the driven element.

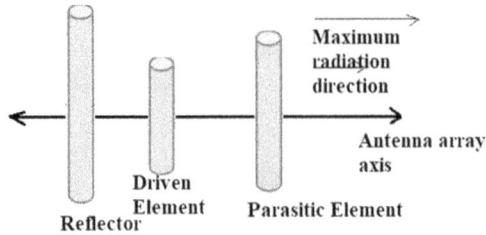

Figure 8.7 Parasitic array.

8.2.6 Advantages of array antenna

1. There is high directivity available. Additionally, the directivity may be altered by choosing the appropriate number of pieces for the circumstance.
2. The conveyed signal's intensity significantly increased.
3. There is an electronic beam guiding option. The beam's direction can therefore be changed from one place to another.
4. There is an increase in signal-to-noise ratio.
5. By giving each element non-uniform input, the radiation pattern may be modified to fit the demands.
6. The design of the antenna array supports better antenna performance.

8.2.7 Disadvantages of array antenna

1. Because there are many antennas set in an antenna array, the construction is massive and takes up a lot of area.
2. Due to the size of the antenna array, maintenance is quite difficult.
3. The array has larger resistive losses.

8.3 WEARABLE ANTENNA

Wearable antennas have generated a lot of interest recently because of their attractive features and promise to enable mobile, wireless communication, and sensing. They must be built of flexible materials and have a low-profile design in order for these antennas to be conformal when put on various body parts. Ultimately, these antennas need to be able to operate with the least degree of degradation while near a person's body. These criteria make designing wearable antennas difficult, especially when taking into account factors like their compact size, the impact of structural deformation and connection to the body, as well as the complexity and precision of manufacturing. The majority of these problems remain in the context of body-worn implementation, notwithstanding minor changes in severity according to applications [8, 9].

A wearable antenna is any antenna that has been specially designed to function while being worn. Examples include smart watches, which frequently have built-in Bluetooth antennas, glasses such as Google Glass, which also have Wi-Fi and GPS antennas, GoPro action cameras, which frequently have Wi-Fi and Bluetooth antennas and are strapped to users to capture their footage, and even the Nike+ Sensor, which pairs with a Smartphone via Bluetooth and is placed in a user's shoe as shown in Figure 8.8. Wearable antennas are rapidly being used by consumer electronics; hence this sectionis devoted to describing the unique difficulties in wearable antenna design.

8.3.1 Challenges of wearable antennas

The design of antennas for wearable technology is particularly challenging due to two issues:

Closeness to the human body: A lossy medium for EMW is the human body. In other words, the body absorbs energy from EMW by converting electric fields into heat. As a result, the antenna effectiveness of your wearable antenna is significantly decreased when an antenna is situated close to the body. For instance, if you construct an antenna and assess its efficiency at −3 dB (50%) it may easily decline to −13 dB (5%) when mounted on the body. The wireless system's performance is severely hampered by this.

Extremely small volumes: The smallest possible wearable technology is required. Everyone dislikes watches with large dipole antennas that protrude from the side. Space is quite limited for wearable technology, especially for items that are worn close to the face (like Google Glass). Because of this, industrial and product designers frequently leave very little room for the antenna, which makes the challenge of antenna design even more challenging.

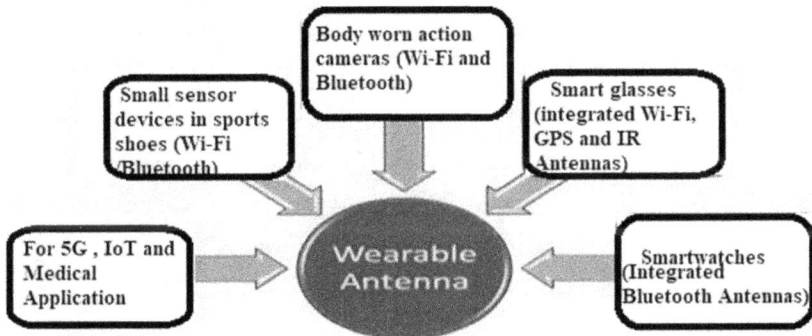

Figure 8.8 Examples of wearable array antenna.

8.3.2 Types of wearable antenna

The wearable antenna can be broken down into three categories. We will explore several antenna types and the impact an absorbing substance put near the antenna can have in order to better understand how the human body influences antennas. We'll start by building three planar antennas out of a sheet of copper tape that is 3.5″ by 1″ in size. In Figure 8.9, the first antenna is a wideband dipole, located on the left. A loop antenna is on the left, and a slot antenna is displayed in the center.

8.3.2.1 Wideband dipole antenna

By taking up larger area, an antenna may be made more broadband, according to the general rule of thumb in antenna design. Consequently, a dipole antenna may be made more broadband by extending the dipole's radius A. Wideband dipoles are another name for these antennas. On 1.5-meter-long

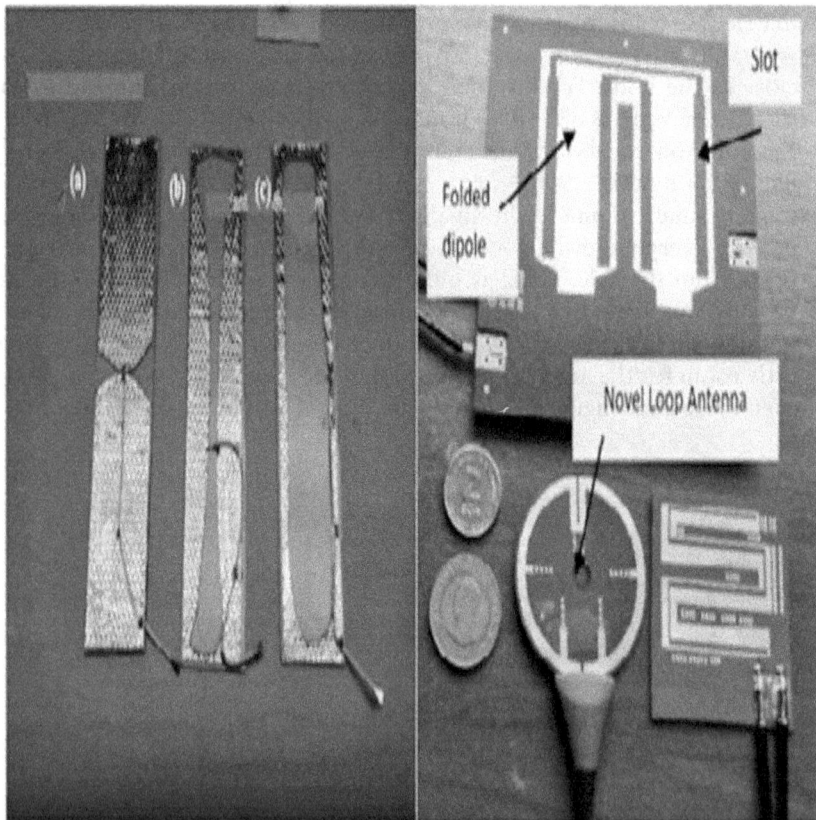

Figure 8.9 Three antennas: (a) wideband antenna, (b) slot antenna (c) loop antenna.

dipoles, for instance, the technique of moment simulations will be carried out. The dipole is half-wavelength long at this length of 100 MHz. Considered are three cases:

At 100 MHz, A = 0.001 m is equal to 1/3000th of a wavelength.
1/100th of a wavelength at 100 MHz is equal to A = 0.015 m.
1/30th of a wavelength at 100 MHz is equal to A = 0.05 m.

8.3.2.2 Slot antenna

Slot antennas are often used for frequencies between 300 MHz and 24 GHz. Slot antennas are frequently employed due to their ability to be cut out of the surface they are to be put on and the fact that their emission patterns are typically omnidirectional (similar to those of a linear wire antenna, as we'll see). Linear polarization characterizes the slot antenna. Size, shape, and what's behind the slot (the hollow) can all be used to modify performance. Assume that a rectangular slit of dimensions a and b have been carved out of an infinite conducting sheet. A slot antenna can be created if certain appropriate fields can be excited in the slot (also known as the aperture), as shown in Figure 8.10.

8.3.2.3 Loop antenna

As seen in Figure 8.11, the little loop antenna is a loop. The loop would be connected in series with the antenna feed points, giving the illusion of a short circuit across the antenna feed when a tiny loop is present. These antennas' low radiation resistance and strong inductive reactance make it challenging to match their impedance to that of a radio (typically 50 Ohms). Due to the fact that impedance mismatch loss may sometimes be more forgivably tolerated in specific systems, these antennas are typically employed as receive antennas. The wavelength (a) is expected to be significantly larger than the radius (a). In the x–y plane, the loop is located.

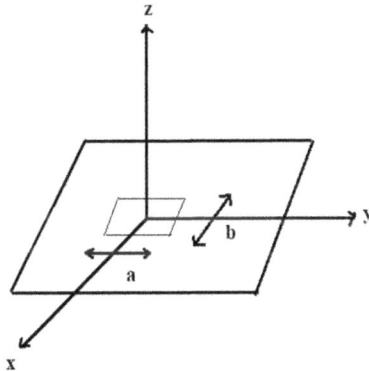

Figure 8.10 Slot antenna.

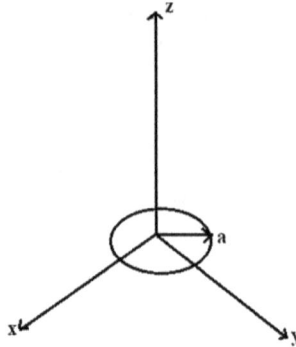

Figure 8.11 Loop antenna.

8.4 ADVANTAGES OF WEARABLE ANTENNA

Even if you are a regular person, utilizing a wearable antenna has several benefits. The wearable antenna also performs well for critical medical and military tasks. Here, we'll start with the benefits of wearing an antenna.

8.4.1 Measures the body's parameters

The wearable antenna's main function is to measure the body's parameters. The antenna is put within your watch to track things like your heart rate and body temperature.

8.4.2 Navigation and tracing

The wearable antenna is practical for military use. The wearable antenna is a crucial application for military operations, whether it be for tracking or navigation.

8.4.3 Public security

For the sake of public security, the wearable antenna is crucial. When a device is linked to the antenna, we can track all other connected devices in addition to humans. Perhaps this explains why the wearable antenna is employed for so many challenging and important tasks.

8.4.4 Versatile and portable

Because most wearable antennas are made to be worn underneath clothing, watches, or sunglasses, they are lightweight and flexible.

Finding crucial information is the main goal of deploying a wearable antenna for crucial military operations. Nobody can predict what will happen during the procedure. Such design principles should be applied to the wearable antenna.

8.4.5 Low costs

The small wearable antenna is employed in military and medical activities. These two uses for the antenna come at a high expense. However, it has become common practice for regular people to monitor their heart rate, blood pressure, etc. The antenna is used by ordinary people under their shirts, watches, sunglasses, and a variety of other items. In order to be accessible to everyone, wearable antennae must become tiny and affordable.

8.4.6 Compact style

The wearable antenna has a small design that makes it convenient to wear everywhere. The antenna is made to be so tiny that it is impossible to think that it might be concealed behind your little watch or slender clothing.

8.5 ISSUES/DISADVANTAGES OF WEARABLE ANTENNA

The wearable antenna has trouble producing satisfactory results for two reasons. To help you comprehend everything, we'll quickly recap those two arguments.

8.5.1 Closeness to the person's body

Since the person's body is susceptible to EMW, it takes advantage of every opportunity to collect energy from the worn antenna. As a result, the antenna results in significant energy losses and performance issues. Due to its proximity to the human body, the wearable antenna has trouble producing accurate data.

8.5.2 Extremely low volumes

Designing a tiny wearable antenna with larger volumes is difficult. For industrial and critical applications, however, purchasing a compact antenna with a bigger capacity is needed. Therefore, creating a high-volume wearable antenna may be difficult for the manufacturer.

8.6 THE APPLICATION OF WEARABLE ANTENNA

Body-worn wearable technologies have attracted a lot of research interest over the past decade or so because they can be used in a variety of

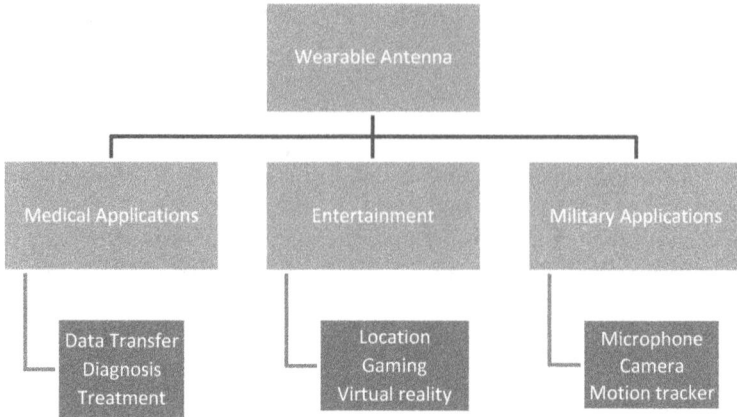

Figure 8.12 Applications of wearable antenna in various fields.

specialized fields that use body-centric communication systems, such as the healthcare industry as a wearable tool to detect critical health problems of the patients, recovery rooms, clinics, operating rooms, homes, and even while on the move. Additionally, data transmission for camera and microphone modules in military applications uses miniaturized antennas [10]. Additionally, these antennas may be utilized for monitoring purposes on children, seniors, and sportsmen. Figure 8.12 illustrates the usual uses of wearable antennas.

8.7 IoT AND MEDICAL APPLICATIONS OF WEARABLE MICROSTRIP ANTENNAS

For sending and receiving data, higher-frequency bandwidth, data rates, etc. are crucial. The compact, higher-frequency wearable antenna is useful for a variety of things. Wearable antennas must be small, light, affordable, and flexible. Slot antennas, PIFA antennas, printed loops, printed dipoles, and microstrip antennas are lightweight, affordable, and small. These antennas might make effective wearable antennas for the Internet of Things (IoT) and biomedical applications.

Printed antennas are frequently low-profile, compact, flexible, lightweight, and less costly than wired antennas. Microstrip antennas can be used to create wearable antennas. Printed antennas have received a lot of attention in the literature over the past couple of decades. The most popular type of printed antenna is a microstrip antenna. However, loop, PIFA, slot, and dipole-printed are all also often used in RF systems. Medical devices, IoT, seekers, and cell phones may all communicate using printed antennas.

MEDICAL

The wearable antenna is crucial to medical procedures that attach to the human body. When a wearable antenna watch is affixed to your skin, it displays your body's present condition. Additionally, several tiny wearable antennas make medical therapy 100% safer.

MILITARY

The wearable antennas for the military are between 15% and 20% smaller than the standard antennas. This is due to the fact that the military should covertly employ antennas to evade adversaries. It goes without saying that the military's wearable antenna helpsto gather crucial information from adversaries. You are already aware that wearable antennas are made to find a person who is hiding a gadget within their clothing. Similarly, it's crucial to know when a commander has visited the enemy's base.

SMART HOME AND SECURITY

Installing the antenna in your home is crucial in ensuring the security of your smart phone. Getting a wearable antenna is another essential component for turning your house into a smart home. The smart home antenna, however, is larger than the wearable and military antenna. As a result, the antenna is more expensive than the little ones.

8.7.1 Design of wearable microstrip antennas

Etched into a low-loss dielectric substrate are microstrip antennas. Figure 8.13 displays a cross-sectional view of the electric fields of a microstrip antenna. Thin conducting patches called microstrip antennas are etched onto substrates

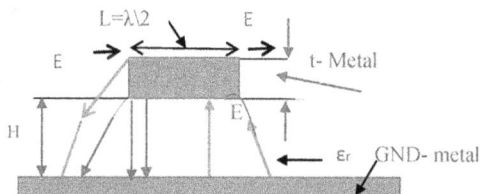

Figure 8.13 A cross-sectional view of microstrip antenna electric fields.

with dielectric constants r and H, where H often falls below 0.1. The presentation includes microstrip antennas. The wearable antenna is either fastened to the wearer's body or placed within a belt.

8.7.2 Benefits of using microstrip antennas

- Low volume and lightweight.
- There is potential for flexible, conformal structures.
- Comparatively inexpensive compared to traditional wired antennas.
- Large uniform arrays and phased arrays are simple to build.
- The effectiveness of these elements is crucial for wearable communication systems.

8.7.3 Microstrip antennas have drawbacks

- Narrow bandwidth (typically 1–5%). Wider bandwidth is, however, feasible with more complicated antenna structures.
- Depending on the thickness of the substrate, low power handling is less than 50 W.
- Limited gain up to 30 dBi, in 16-16 arrays.
- High-frequency feed network losses exceeding 12 GHz.

Over the past few decades, research on wearable antennas has become increasingly important because of the critical role that these antennas will play in the 5G communication system for wearable applications, IoT, and biomedical systems. These antennas have unrivaled qualities such as being lightweight, made from inexpensive flexible substrates, requiring straightforward fabrication methods, and being portable. Wearable antennas operate in a particular environment, such as in close proximity to the human body, and thus their design concerns and specifications must be clear. The materials for antenna design are chosen based on a number of factors, such as their resistance to mechanical deformation caused by bending, stretching, and cramping, as well as their ability to survive a variety of meteorological conditions (rain, dust, sunshine, high temperatures, etc.). The functioning frequency bands are the only ones where wearable antennas may be used. Various wearable antennas that are resonant in the VHF, UHF, and microwave frequency ranges are shown in this context for off-body, on-body, and in-body communication. On the basis of the chosen flexible materials, the fabrication process of wearable antennas is outlined, along with the benefits and drawbacks of each material. For these kinds of antennas, manufacturing processes including line patterning, wet etching, inkjet printing, screen printing, and embroidery are frequently used. The main needs for wearable antennas in wireless communication, their applications in biomedicine, and design criteria, as well as operating frequency ranges, manufacturing procedures, and measurement methodologies, are highlighted in this chapter. The latest advancements in wearable textile/non-textile antennas for WBAN are reviewed in detail here.

8.8 BIOMEDICAL APPLICATION OF WEARABLE ANTENNA

There has been a lot of attention recently on wearable flexible/semi-flexible devices for biomedical telemetry applications. For the development of wearable nodes, using flexible or semi-flexible materials is getting a lot of interest [11]. The development of implanted medical devices for use in endoscopy [12, 13], brain recording [14], glucose monitoring [15, 16], and intracranial pressure monitoring [17] are among the examples of advances in wearable technology.

To decrease the size of wearable technology and inefficient power use, a flexible wearable antenna design is crucial. This chapter presents a unique Ultra Wide band antenna for the detection of different tumors in the body utilizing microwave imaging. Jeans cloth measuring $25 \times 24 \times 1.6$ mm^3 with slots in the patch is used to create the antenna. The results are compared after the antenna is tested and simulated on various body phantoms. As the device responsible for receiving and transferring signals between the implanted device and the wearable network, the wearable antenna is a crucial part of the wearable network [18, 19]. Because the human body is a lossy platform for EM waves and a lot of EM waves are absorbed by the body in the forms of heat and energy, the antenna should be made as efficient as possible [20]. In order to minimize the number of backward radiations (the Specific Absorption Rate, or SAR) that might harm human tissues, a specific design consideration is needed for the antenna [21]. The total size of the antenna must also be taken into consideration [22]. The antenna's total size must be as minimal as feasible.

At the end of the nineteenth century, EMWbegan to be used for medical purposes in the form of X-rays. Since then, EM waves have developed, along with fresh difficulties in the medical industry. Biotelemetry, biomedical treatment, and diagnostics have benefited greatly from EMW [23, 24]. Antenna implantation inside the human body is necessary for biomedical applications that use EMW. The implanted device in the body of the patient is designed to gather the patient's information and to transmit it wirelessly to the base station. The use of pacemakers and tablets with sensing capabilities as implanted devices first gained popularity in the 1960s. The development of implantable technology presents new difficulties for biomedical applications. These surgically placed gadgets gather patient data and transmit it wirelessly to the base station. The present focus of research is to expand the scope of these biomedical applications for ongoing patient monitoring [25, 26].

In the person's body, wearable antennas can be used in two different biological scenarios: biomedical treatment and biotelemetry. A wireless communication channel between the human body and the outside world can be created through biotelemetry. Diseases are treated and numerous physiological indicators are monitored as part of biomedical therapy and diagnostics. Because the patient spends less time in the hospital, these applications lower the cost of healthcare. The healthcare monitoring system, known as Wireless Body Area Network (WBAN), employs implanted technology inside the human body. The patient's home healthcare is monitored via this system. Amodel of the WBAN system is shown in Figure 8.14.

In spite of the fact that there is a lot of academic and commercial interest in the subject, not many applications utilize this technology. Along with the pacemaker, the condition of the Glucoday and Pillcam devices for glucose monitoring and endoscopy, respectively, might be viewed as examples of commercially accessible implanted sensors in their respective fields. The former is a wearable device with a widespread microfiber that can measure the glucose concentration from the interstitial fluids, as shown in Figure 8.15(a). The latter, seen in Figure 8.15(b), transmits pictures from the digestive tract across a relatively small distance (patient skin is stitched with receivers). Although these technologies substantially enhance medical diagnosis, they do not satisfy the demand for a patient-centered healthcare monitoring system that is as "transparent" as feasible.

8.9 OBJECTIVES OF WEARABLE ANTENNA IN HEALTHCARE MONITOR

- To transmit accurate data from inside the human body to a remote base station (BS).
- A WBAN system consists of a number of components. Figure 8.16 displays the various WBAN system units.
- Proper coordination between all the components is necessary for the system to operate as intended.
- The most crucial factor before a WBAN system is put into actual use is adequate testing.

8.10 WIRELESS BIOTELEMETRY

The smart items that surround us are wirelessly connected in various ways in the new IoT age. This paradigm, in which information and communication systems are developing into devices embedded in the environment, is dominated by sensor network technologies and Radio Frequency Identification (RFID) [27]. WBANs support the IoT's integration into the healthcare sector.

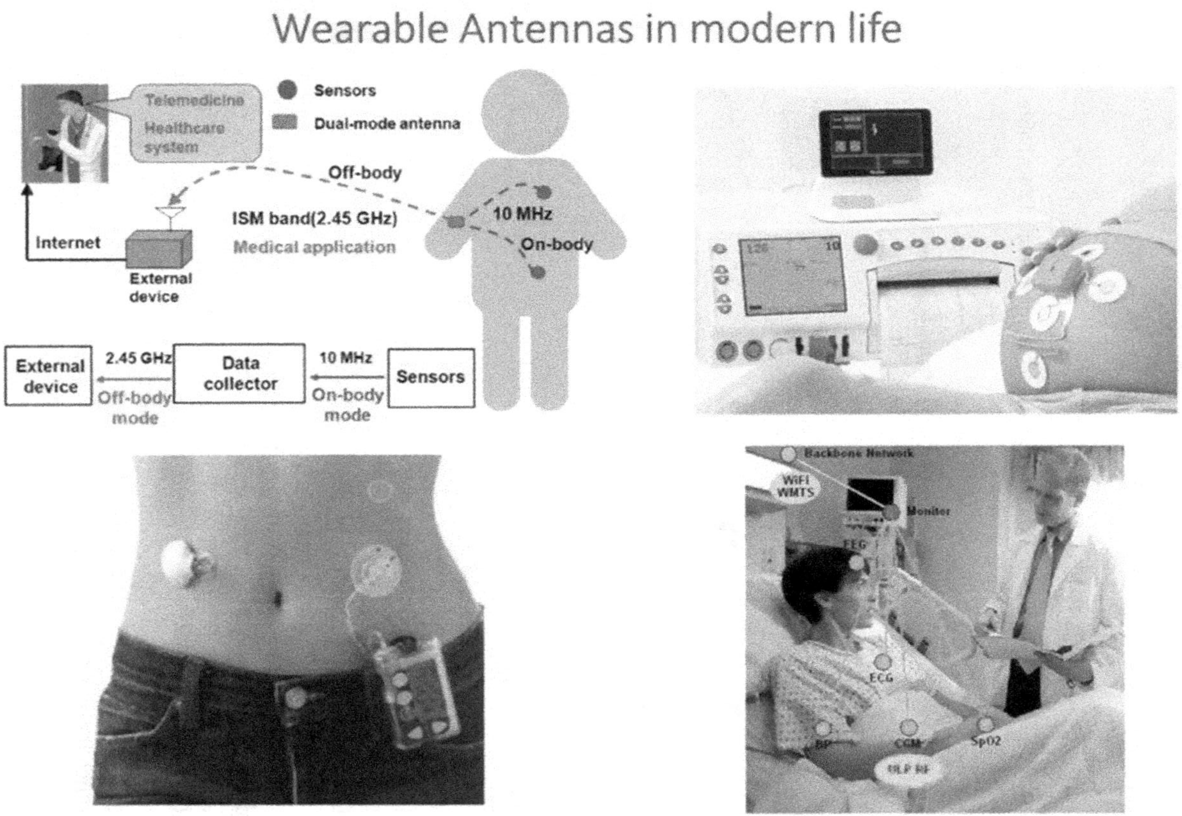

Figure 8.14 Wearable antenna model in modern life.

Figure 8.15 (a) Glucoday, (b) Pillcam; examples of commercially available implantable monitoring devices.

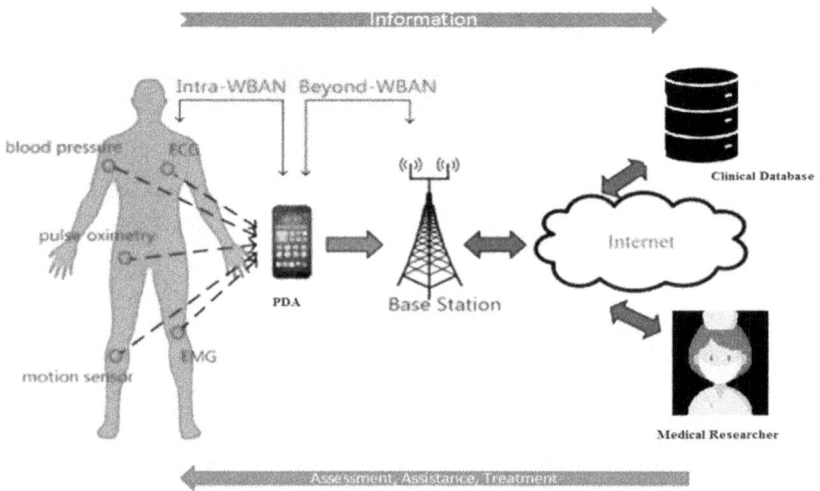

Figure 8.16 Wireless Body Area Network (WBAN).

This network continuously receives large amounts of data from which microprocessors must extract the pertinent details and properties. Data collection using a hierarchy is also employed. As a result, the network's next node in the dataset has many sensor nodes, each of which processes data locally. The gathering unit is in charge of the composition of the data. These collectors combine the information from the nodes by acting as a liaison between higher levels of the network and the nodes. The information gained at each higher level can serve as feedback to the lower levels, helping to enhance feature extraction, categorization, and sensor coordination [28, 29].

8.10.1 Base station

A base station is made up of many supporting systems. The control module, receiving system, and internet modem are among the subsystems. The information is stored in the control module, which uses it to operate the entire system. Using an antenna, the receiver module collects the data and transmits it to the control module. The modules can be connected to one another utilizing wireless data transmission with the aid of an internet modem.

8.10.2 Channel

Analysis of scattering and multipath propagation is required for EM wave propagation between the base station and the implanted device. The effectiveness of the propagation channel can be increased by minimizing the aforementioned impacts.

8.10.3 Person's body

The person's body is a challenging situation for a fixed antenna due to its extremely complex, lossy, and dispersive nature. The properties of the surrounding environment are crucial for creating an implanted antenna. The many steps of implantation include analysis, design, realization, and characterization. The person's body must be taken into account at every level as the surrounding environment.

8.10.4 Insulations and packing

The close materials used in the antenna's substrate and superstrate change how it behaves. Unavoidable requirement for in-body technologies to prevent negative impacts on biological tissues.

8.10.5 Electricity and electronics

In addition to the antenna, the biomedical device's overall volume is also influenced by its electronic parts and power source. The device's electronic parts combine the communication and processing units and ensure correct operation. The power supply's purpose is to deliver the necessary power as determined by these module components.

8.10.6 Wearable antennas (Wireless Body Area Networks)

One important component of a WBAN network is an implantable antenna. The compactness, price, and efficiency of the radiator are the main considerations. While building an antenna it is important to consider factors, including radiation efficiency, band breadth, and interaction with biological tissues that are prone to loss. Figure 8.16 illustrates the two layers of the WBAN: The "Intra-BAN" is the top layer. This tier's communication is conducted between a controller node and sensor nodes. "Beyond-BAN" is the third layer.

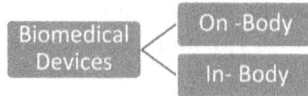

Figure 8.17 Categories of biomedical devices.

8.11 CATEGORIES OF BIOMEDICAL DEVICES

Depending on where they are, biomedical devices might be either on the body or within the body. The decision to use one type of device over another may be influenced by economic factors, societal changes, technology advancements, and research developments. Wearable or on-body technology is already a part of our daily lives. For instance, sensing wristbands and smartwatches are already widely available. Other product categories, such as neckwear and jewelry for fashion, headgear or eyeglasses for virtual or augmented reality, smart shoes and body wear for sports and fitness, etc., will flourish as a result of new trends in wearable technology. According to [30], by 2026 the market for these smart wearable products will move over $150 billion annually.

8.11.1 Wireless on-body or wearable gadgets

Wireless on-body or wearable gadgets are being implemented in the health-care system because they allow for the non-invasive monitoring of vital signs. Miniaturization, security, standardization, energy efficiency, robustness, and unobtrusiveness are some of the issues that have still to be overcome.

8.11.2 Devices utilized in the body

The majority of these devices are still under development and need a special prescription from a doctor in order to be utilized. Such devices first became available when Rune Elmgvist and his colleagues were successful in placing the first pacemaker inside a human body in 1958. Since then, in-body devices—often referred to as Implantable Medical Devices (IMD)—have filled a void where physiological parameters are only available from within the human body or when wearable technologies are insufficiently potent. However, there arenot yet many in-body medical deviceson the market; the majority are still in development, at the testing or prototype stage, with the exception of implanted pacemakers and a few ingestible endoscopic capsules with biotelemetry capabilities.

Depending on how they are inserted into the human body, in-body wireless devices are categorized as implantable, indigestible, or injectable. In the case of wearables, however, there are still more obstacles to overcome, including a higher degree of concerns, charging, and other issues.

8.12 CONCLUSIONS

To expand the breadth of biomedical applications, research in several domains is possible using the WBAN. Optimizing the implanted devices is the main difficulty facing researchers working on microwave antennas. Getting to the optimal circumstances is the goal communication network. The implanted antennas, one of the most difficult components of a WBAN system, have received the majority of attention while taking into consideration a wide variety of technological and medical problems. Along with the more traditional antenna difficulties, such as miniaturization and the resulting reduction in radiation efficiency, the interaction of the electromagnetic field with biological tissues that are prone to loss must also be taken into consideration. Additionally, biocompatibility, physical requirements, and the intimate cohabitation of all the components present in the total system. Furthermore, consideration must be given to biocompatibility, physical requirements, and the close cohabitation of all the components included in the overall WBAN system.

REFERENCES

1. A. Rosen, M.A. Stuchly, and A. Vander Vorst, "Applications of RF/Microwaves in Medicine," *IEEE Trans. Microw. Theory Tech.*, vol. 50, no. 3, pp. 963–974, 2002.
2. A. Polemi, A. Toccafondi and S. Maci, "High-frequency green's function for a semi-infinite array of electric dipoles on a grounded slab—Part I: Formulation", *IEEE Trans. Antennas Propag.*, vol. 49, no. 12, pp. 1667–1677, Dec. 2001.
3. D. Panescu, "Emerging Technologies [Wireless Communication Systems for Implantable Medical Devices]," *IEEE Eng. Med. Biol. Mag.*, vol. 27, no. 2, pp. 96–101, 2008.
4. R. Bashirullah, "Wireless Implants," *IEEE Microwave. Mag.*, vol. 11, pp. 345–351, 2010.
5. R.S. Mackay, "Radio Telemetering from Within the Body: Inside Information Is Revealed by Tiny Transmitters that Can Be Swallowed or Implanted in Man or Animal," *Science*, vol. 134, no. 3486, pp. 1196–1202, 1961.
6. N. Singh, "IOT Enabled Hybrid Model with Learning Ability for E-Health Care Systems", *Measurement: Sensors*, vol. 24, pp. 1–14, 2022. https://doi.org/10.1016/j.measen.2022.100567
7. P. Valdastri, A. Menciassi, A. Arena, C. Caccamo, and P. Dario, "An Implantable Telemetry Platform System for in vivo Monitoring of Physiological Parameters," *IEEE Trans. Inf. Technol. Biomed.*, vol. 8, no. 3, pp. 271–278, 2004.
8. K. Najafi, "Packaging of Implantable Microsystems," *Proc. IEEE Sensors*, vol. 21, no. 4, pp. 58–63, 2007.
9. E.A. Johannessen, L. Wang, C. Wyse, D.R.S. Cumming, and J.M. Cooper, "Biocompatibility of a Lab-on-a-Pill Sensor in Artificial Gastrointestinal Environments," *IEEE Transition Biomed. Eng.*, vol. 53, no. 11, pp. 2333–2340, 2006.

10. D. Halperin, T. Kohno, T.S. Heydt-Benjamin, K. Fu, and W.H. Maisel, "Security and Privacy for Implantable Medical Devices," *IEEE Pervasive Comput.*, vol. 7, no. 1, pp. 30–39, 2008.

11. A. Cucini, M. Albani and S. Maci, "Truncated floquet wave full-wave analysis of large phased arrays of open-ended waveguides with a nonuniform amplitude excitation," *IEEE Trans. Antennas Propag.*, vol. 51, no. 6, pp. 1386–1394, Jun. 2003.

12. R. Kindt, K. Sertel, E. Topsakal and J. Volakis, "Array decomposition method for the accurate analysis of finite arrays," *IEEE Trans. Antennas Propag.*, vol. 51, no. 6, pp. 1364–1372, Jun. 2003.

13. D. Zarifi, A. Farahbakhsh, A. U. Zaman and P.-S. Kildal, "Design and fabrication of a high-gain 60-GHz corrugated slot antenna array with ridge gap waveguide distribution layer," *IEEE Trans. Antennas Propag.*, vol. 64, no. 7, pp. 2905–2913, Jul. 2016.

14. M. Sharifi Sorkherizi, A. Dadgarpour and A. A. Kishk, "Planar high-efficiency antenna array using new printed ridge gap waveguide technology," *IEEE Trans. Antennas Propag.*, vol. 65, no. 7, pp. 3772–3776, Jul. 2017.

15. Y. Li and K.-M. Luk, "60-GHz dual-polarized two-dimensional switch-beam wideband antenna array of aperture-coupled magneto-electric dipoles," *IEEE Trans. Antennas Propag.*, vol. 64, no. 2, pp. 554–563, Feb. 2016.

16. W. Xia, K. Saito, M. Takahashi, and K. Ito, "Performances of an Implanted Cavity Slot Antenna Embedded in the Human Arm," *IEEE Trans. Antennas Propag.*, vol. 57, no. 4, pp. 894–899, 2009.

17. L. Bolomey, Generic DSP-based implantable body sensor node. Ph. D. dissertation, EPFL, Lausanne, 2010.

18. J. Kim, and Y. Rahmat-Samii, "Implanted Antennas Inside a Human Body: Simulations, Designs, and Characterizations," *IEEE Trans. Microwave Theory Tech.*, vol. 52, no. 8, pp. 1934–1943, 2004.

19. R.W.P. King, and G.S. Smith, *Antennas in Matter: Fundamentals, Theory, and Applications*, 1st ed. Cambridge, MA; London, England: The MIT Press, 1981.

20. C. Gabriel, *Compilation of the Dielectric Properties of Body Tissues at RF and Microwave Frequencies*. Brooks Air Force Base, Texas (USA), Tech. Rep. Report N.AL/OE-TR, 1996.

21. N. Singh, G. K. Prajapati, "Wireless Sensor Network Security Problems-A Review", *International Journal of Emerging Technology and Advanced Engineering*, vol. 8, no. 1, 2018.

22. P.S. Hall, and Y. Hao, *Antennas and Propagation for Body-centric Wireless Communications*, Norwood, MA: Artech House. Chapter 9, 2009.

23. R. Fenwick, and W. Weeks, "Submerged Antenna Characteristics," *IEEE Trans. Antennas Propag.*, vol. 11, no. 3, pp. 296–305, 1963.

24. R. Hansen, "Radiation and Reception with Buried and Submerged Antennas," *IEEE Trans. Antennas Propag.*, vol. 11, no. 3, pp. 207–216, 1963.

25. R. Moore, "Effects of a Surrounding Conducting Medium on Antenna Analysis," *IEEE Trans. Antennas Propag.*, vol. 11, no. 3, pp. 216–225, 1963.

26. P.B. Johnson, S.R. Whalen, M. Wayson, B. Juneja, C. Lee, and W.E. Bolch, "Hybrid Patient Dependent Phantoms Covering Statistical Distributions of Body Morphometry in the U.S. Adult and Pediatric Population," *Proc. IEEE*, vol. 97, no. 12, pp. 2060–2075, 2009.

27. P. Soontornpipit, C. Furse, and Y.C. Chung, "Design of Implantable Microstrip Antenna for Communication with Medical Implants," *IEEE Trans. Microwave Theory Tech.*, vol. 52, no. 8, pp. 1944–1951, 2004.

28. P. Soontornpipit, C.M. Furse, and Y.C. Chung, "Miniaturized Biocompatible Microstrip Antenna Using Genetic Algorithm," *IEEE Trans. Antennas Propag.*, vol. 53, no. 6, pp. 1939–1945, 2005.

29. J. Kim, and Y. Rahmat-Samii, "Planar Inverted-F Antennas on Implantable Medical Devices: Meandered Type Versus Spiral Type," *Microw. Opt. Technol. Lett.*, vol. 48, no. 3, pp. 567–572, 2006.

30. W.-C. Liu, S.-H. Chen, and C.-M. Wu, "Implantable Broadband Circular Stacked PIfA Antenna for Biotelemetry Communication," *J. Electromagn. Waves Appl.*, vol. 22, no. 13, pp. 1791–1800, 2008.

Chapter 9

Circular shaped 1×2 and 1×4 microstrip patch antenna array for 5G Wi-Fi network

Rovin Tiwari and Raghavendra Sharma
Amity University, Gwalior, India

Rahul Dubey
MITS, Gwalior, India

9.1 INTRODUCTION

Advanced wireless communication stands as a cornerstone of our modern interconnected world, driving innovation, enabling new technologies, and revolutionizing the ways in which we interact, work, and live [1]. The fifth generation of wireless communication, 5G, is a game-changer in the field. It promises significantly higher data rates, ultra-low latency, massive device connectivity, and network slicing. 5G enables the seamless integration of diverse applications such as augmented reality, virtual reality, autonomous vehicles, remote surgery, and smart cities.

Figure 9.1 presents the new Wi-Fi channel name under the 5G network. Wireless communication technologies have become an integral part of modern life, transforming industries, enabling new business models, and empowering individuals across the globe. From the simple voice calls of 1G to the immersive experiences of 5G and beyond, the journey of wireless communication continues to shape the way we live, work, and interact [2]. Microstrip patch antennas have gained significant attention and utilization in the context of 5G wireless communication due to their compact size, the ease of manufacture, and compatibility with modern electronics. These antennas offer a range of advantages that make them well-suited for 5G applications.

When organized into an array configuration, these antennas hold the potential to not only address the challenges of 5G and Wi-Fi networks but also unlock new capabilities that can transform our digital experiences [3]. This innovative approach offers advantages such as beam forming, improved signal strength, interference mitigation, and enhanced coverage, making it an ideal candidate for supporting the data-hungry applications of today's world. In this exploration, research delves into the realm of MPA-A technology for 5G Wi-Fi networks [4]. It delves into the fundamental principles that underlie microstrip patch antennas and the benefits they bring to wireless communication. Additionally, we will examine how the

DOI: 10.1201/9781003422440-9

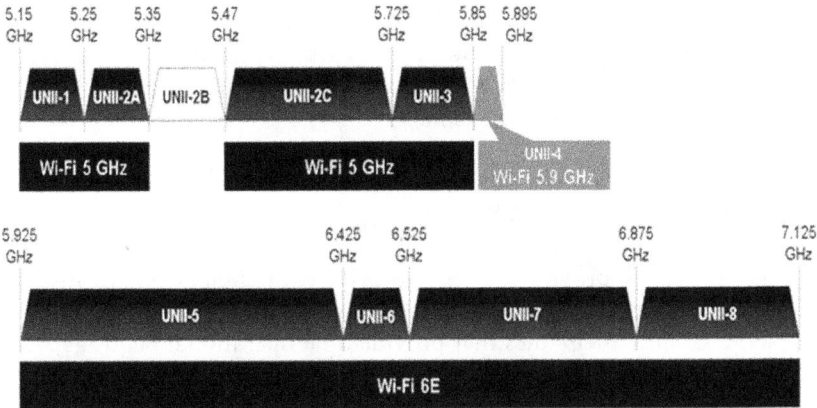

Figure 9.1 5G Wi-Fi New channel.

deployment of antenna arrays can revolutionize the way we experience connectivity, enabling seamless data transfer, reduced latency, and support for an ever-growing number of devices. By understanding the intricacies and potential of this technology, we can grasp its role in shaping the future of wireless communication and how it aligns with the demands of the 5G and Wi-Fi landscape [5].

Due to their relatively high absorption, the wavebands employed by Wi-Fi are best suited for direct line-of-sight communications. Common obstructions such as walls, pillars, domestic appliances, etc., may significantly restrict range; however, this reduces interference across networks in crowded places [6].

A circular MPA-A is a collection of circular microstrip patch antennas arranged in a specific pattern to achieve a desired radiation pattern. A circular MPA consists of a circular metallic patch on top and ground plane, a middle layer made from the dielectric substrate [7]. The patch is excited by a coaxial probe or a microstrip line and radiates electromagnetic waves in the desired direction [8].

An antenna array is a collection of two or more antennas that work together to achieve a specific radiation pattern. By carefully arranging the antennas in the array, it is possible to steer the radiation pattern in a particular direction or shape. Antenna arrays are used in various applications, such as radar, wireless, and communication with satellite systems [9].

The circular MPA-A can be designed using various techniques, such as the linear, planar, or circular array. In a linear array, the antennas are arranged in a straight line, whereas in a planar array, the antennas are arranged in a two-dimensional plane. In a circular array, the antennas are arranged in a circular pattern [10].

The remaining chapter consists of four sections. Part 9.2 provides the literature background to MPA-A. Following this, part 9.3 presents the methodology and part 9.4 shows the work of simulation, experiment, and result analysis. Finally, part 9.5 offers the conclusion.

9.2 LITERATURE SURVEY

M. Alsukour et al. provide a side-edge MPA with circular polarization for 5G wireless networks and smartphones. The planned antenna is made up of a series of microstrip lines that provide dual operational bands at 5 & 6 GHz [1].

S. P. Pavithra et al. report the design of a high-gain circular and rectangular MPA. When comparing the gain of a circular MPA designed using cutting-edge antenna design and FEKO software to that of a rectangular patch antenna, the former has a 10 dB advantage [2].

S. Jabeen et al. present 6–75 GHz (Wi-Fi-6E) ranges used by the monopole slot antenna. The total volume of the design structure is 103.5 mm × 103.5 mm × 0.51 mm. Minimum isolation is more than −22 dB, and the highest envelop correlation coefficient is 0.124 [3]. Y. M. Al Hyasat et al. describe a novel two-element ultra-wideband MPA intended to service 5G communications, Wi-Fi-5, and Wi-Fi-6. The antenna has an operational range of 2.23 GHz to 12.57 GHz [4].

Y. M. Hyasat presents a two-element wideband MIMO antenna with an FBW of 97.5%, operating in the frequency range. The antenna consists of a pair of parallel rectangular patches, each with a circular slot, with E placed within the slot [5]. N. Praween et al. describe a dual-band multilayer MPA made from LTCC and FR-4, Wi-Fi and sub-6 GHz for 5G use scenarios for the impedance BW S11-10 dB, which covers 34% of the C-Band [6].

M. A. Jiddney et al. considers a patch in the form of a circle. The planned antenna operates across the 24–31 GHz, or +10 dB, portion of the 5G spectrum. A gain of up to 5 dB is attained with an efficiency of roughly 60%. The developed antenna is intended for use in advanced 5G systems [7].

To facilitate 5G and wireless communication, F. C. Gül et al. report the results of their research into, and development of, a wideband circular patch antenna. The scattering parameters, radiation patterns, and gain values are all presented, along with simulation and experiment results [8].

A. Assoumana et al. used a DGS for a size reduction of between 50 and 70% from a conventional 3.6 GHz circular patch antenna. Area reductions of 75% and 91% were thus accomplished. The benefits obtained include: The normal antenna has a dBi of 8.69, whereas the reduced-area antennas have dBi values of 5.17 and 2.41, respectively [9].

A 4×4 MIMO antenna for tiny cells is given by C.-C. Chang et al. The suggested antenna is compatible with the 5G (3300–5000 MHz). The vertical set-up, which consists of a PIFA and a monopole antenna, operates in the Wi-Fi frequency range [10].

An antenna for 5G and WiFi-5/6 networks is presented by S. C. Das et al. To minimize its visual impact, the proposed antenna has been miniaturized. It now partly covers the S and C bands, and its BW has been increased its directivity is 3.368 dB, while its VSWR is 1.017 and 1.025, respectively [11].

According to R. A. Panda et al., circular patches may be seen on the substrate's surface. The cylindrical probe, which has a higher resonant-frequency, has been used to complete the modification and feeding [12]. N. P. Kulkarni et al. for use in a 5G Wi-Fi network operates at 5.9 GHz, yielding broadband with a fractional BW of 95% [13].

S. Khade et al. suggest designing the specified antenna both with and without DGS compared at length. Resonance is achieved at 3.89 GHz and 2.9 GHz in the DGS design, which allows for operation throughout a frequency range of 2.22 GHz to 4.87 GHz. In addition, a frequency range of 8.9 GHz to 9.8 GHz is realized [14]. To make it suitable for 5G usage, J. Colaco et al. study the design of a circular MPA with a resonance frequency of 28.4 GHz and a microstrip feed line. [15].

S. R. Govindarajulu et al. introduce the integration of nearby parts, and a Wilkinson power divider is used due to the huge available connection size. The array has excellent matching between 5.85 and 6.48 GHz, as shown by simulation, and between 17.29 and 29 GHz. In all frequency ranges, the actual gain is quite close to the theoretical maximum [16]. A circular printed monopole antenna with two frequency bands is offered for use in wireless fidelity communication by N. Cascalho et al. There is a presentation of configurations, simulations, and outcomes [17].

The MPA developed by D. Sharma et al. may be used in WLAN, Wi-Max, and DSRC networks. At 5.9 GHz, the antenna has a gain of 7.6 dBi and a Return-loss of 48.47 dB. Gain, VSWR, and Return-loss have all been enhanced in comparison to a standard MPA operating at 5.9 GHz thanks to the use of a metal grating [18].

S. X. Ta et al. show analysis of the design shows that this approach is very effective and reliable for 5G applications, broadband characteristics, and a unidirectional radiation pattern over three frequency bands [19].

R. Tiwari et al. present BW optimization and the design of a 22 microstrip patch array antenna by an effective machine learning approach. The antenna's patch and top are both made of copper, while the bottom is made of FR-4 Epoxy substrate. Antenna parameters such as BW are optimized with the help of the random forest regressor machine learning approach [20].

R. Tiwari et al. demonstrate how to increase the MPA's performance in terms of accuracy, MMSE, and MAE. With this data-driven simulation approach, the antenna optimization process might potentially be sped up significantly. The benefits of evolutionary learning and dimensionality reduction techniques are also explored, as are their applications in analyzing antenna performance [21].

9.3 METHODOLOGY

The design steps for a circular MPA-A can vary depending on the specific requirements and the specifications of the application. However, the following steps provide a general guideline for designing a circular MPA-A:

Determine the frequency range: The frequency range of the antenna array should be determined based on the frequency band of the application. In the case of a 5G Wi-Fi network, the sub-6 GHz frequency band is typically used.

Choose the antenna element: A circular microstrip patch antenna is a good choice for the antenna element due to its circular polarization, low profile, and ease of integration. The antenna dimensions can be determined based on the desired operating frequency and radiation characteristics.

Determine the antenna array configuration: The configuration of the antenna array should be selected based on the desired radiation pattern and gain requirements. Linear, circular, and planar arrays are common configurations for circular microstrip patch antenna arrays.

Determine the feeding network: The feeding network should be designed to ensure proper phase and amplitude control of the signals. Various feeding techniques can be used such as coaxial, microstrip, or aperture-coupled feeding.

Simulate the antenna array using electromagnetic simulation software: The antenna array should be simulated using simulation software based on EM to optimize the design parameters and verify the radiation characteristics. The simulation results can be used to fine-tune the antenna dimensions, feeding network, and antenna array configuration.

Fabricate and test the antenna array: After the simulation, the antenna array should be fabricated using a PCB or another suitable substrate. The antenna array should then be tested to verify the radiation characteristics and performance.

Tune and optimize the antenna array: The performance of the antenna array can be further optimized by fine-tuning. This process may involve adjusting the dimensions of the antenna element, feeding network, or antenna array configuration.

A circular microstrip patch antenna array involves determining the frequency range, selecting the antenna element, choosing the antenna array configuration, designing the network of feeding, simulating the antenna array, and fabricating and testing the antenna array. Proper design and optimization can result in a compact and high-performance antenna system that can improve the coverage and capacity of a 5G Wi-Fi network.

For the calculation of the width and length of the radiating patch, we need to choose the following parameters [5].

The design equations are as follows:

9.4 RESONANT FREQUENCY OF THE CIRCULAR MICROSTRIP PATCH ANTENNA

$$f = (c/2\pi) * \mathrm{sqrt}(\varepsilon_{\mathrm{eff}})/(d+h) \tag{9.1}$$

Here, f is the resonant $_{\mathrm{frequency}}$, c is the speed of light in free space, $\varepsilon_{\mathrm{eff}}$ is the substrate material effective dielectric constant, d is circular patch diameter, and h is the substrate height.

9.5 THE INPUT IMPEDANCE OF THE CIRCULAR MPA

$$Z_{\mathrm{in}} = (Z_0/2\pi) * \ln(1 + 2\varepsilon_r/(\varepsilon_r + 1) * (d/h)) \tag{9.2}$$

where Z_0 is the feed line characteristic impedance, ε_r is the substrate material relative dielectric constant.

9.6 ANTENNA ARRAY FACTOR FOR A CIRCULAR MPA-A

$$AF(\theta,\varphi) = (1/N) * \mathrm{abs}(\mathrm{sum}(\cos(k * d * \mathrm{in}(\theta) * \cos(\varphi) + k * d * \tag{9.3}$$
$$\times \sin(\theta) * \sin(\varphi)) * \exp(-j * k * r) * AF_{\mathrm{psi}})) * *$$

where $AF(\theta, \varphi)$ is the antenna array factor at a given direction, N is the number of antenna elements in the array, k is the wave number, d is the spacing between the antenna elements, θ is the elevation angle, φ is the azimuth angle, r is the distance from the antenna array, and AF_{psi} is the array factor of a single antenna element in the array.

These equations provide a starting point for the design of a circular MPA-A, but the actual design process involves simulation and optimization to achieve the desired radiation characteristics and performance.

9.7 RESULTS AND ANALYSIS

The CST microwave studio is used to carry out the simulation. The following are the dimensions, in addition to the other constraints: The lower frequency, denoted by fL, is 4 GHz, and the higher frequency, denoted by fH, is 8 GHz. The FR4 material has a dielectric constant of 4.4, denoted by the symbol r. The height of the patch is 0.035 mm, and the length and width of the patch are 20.5 mm and 15 mm respectively. The length and width of the ground are 3 mm and 12 mm respectively, while the height of the ground is 0.035 mm. The feeding patch has dimensions of 8 millimeters by 2 millimeters.

9.7.1 1×2 MPA array design

The circular form microstrip patch antenna array that has partial grounding is the foundation of the design that has been presented in Figure 9.2. A microstrip-fed patch is employed, and the edge of the microstrip patch is linked directly to a conducting strip shown in Table 9.1.

Figure 9.3 depicts the suggested design from both the front and the rear of the antenna hardware. The double-sided PCB is used to make the antenna. The connection known as the Sub Miniature version A (SMA) is attached in the role of a port."

The value of S11 is shown to be −59.96 dB for Band-I in Figure 9.4, which corresponds to the resonance frequency of 5.4 GHz.

The value of S11 or return-loss is shown to be −34.18 dB for Band-II in Figure 9.5 when measured at the resonant frequency of 6.4 GHz.

Figure 9.6 show the suggested antenna has a BW of 916 MHz, which covers a frequency range of 5.8 GHz to 4.9 GHz.

Figure 9.7 shows the suggested antenna has a BW of 2012 MHz, which corresponds to the frequency range of 7.8 GHz to 5.8 GHz.

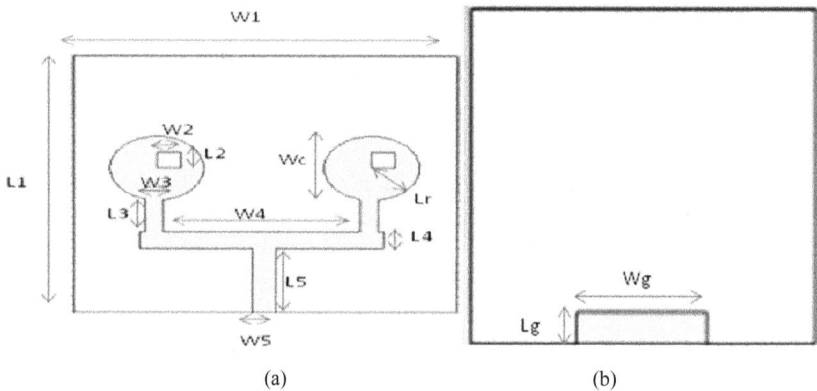

(a)   (b)

Figure 9.2 Antenna view of 1×2 (a) top (b) bottom.

Table 9.1 The proposed antenna dimensions (in mm)

Length	Dimension	Width	Dimension
L1	32	W1	32
L2	2	W2	2
L3	4.13	W3	1.5
L4	2	W4	16.5
L5	8	W5	2
Lr	4	Wc	27.6
Lg	3	Wg	12

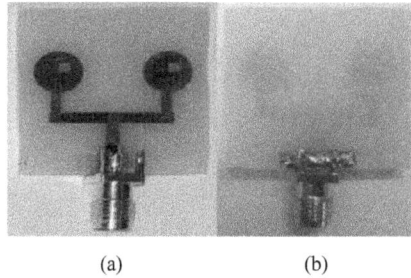

(a)   (b)

Figure 9.3 Hardware view of 1×2 (a) top (b) bottom.

Figure 9.8 demonstrates the antenna hardware testing in the vector network analyzer. This antenna hardware provides results shown in Table 9.2.

9.7.2 1×4 MPA array design

Figure 9.9 shows the extension of the 1×2 antenna array as a 1×4 antenna array with a single cut element is 0.75 mm × 0.75 mm and the overall dimension is the same.

Figure 9.10 presents the partial ground as the same as 1×2. The dimension of a fed patch is 6 mm × 2 mm. Other antenna dimensions used shown in Table 9.3.

Figure 9.11 shows the simulation in all the coordinate systems in the CST plan, electric field, magnetic field, and wave radiation in three different directions.

For the antenna to work correctly, the value of the return loss must be lower than −10 dB. The value of S11 at the resonance frequency of 4.5 GHz is shown to be −30.18 dB in Figure 9.12.

According to Figure 9.13, the optimal frequency band for the 1×4 antenna array has a BW of 843 MHz, which ranges from 5 GHz to 4.16 GHz.

Figure 9.14 presents the designed antenna array power. The total power going in to the antenna is 0.5 W. The radiated power is 0.49 W, so therefore the loss is 0.01 W, meaning that the antenna array power loss is very minimal. The total radiated power is 30 dBmW by the proposed antennas.

The axial ratio vs. theta/degree of an MPA array with the partial ground is shown in Figure 9.15. The axial ratio may be used to determine whether or not an antenna has circular or linear polarization; if the axial ratio of the antenna is up to 3 dB, then the antenna behaves like it has circular polarization. Wi-Fi antennas always have a vertical orientation and a circular polarization pattern. The antenna sensitivity is −40 dBm to −80 dBm for the Wi-Fi router application.

Figure 9.16 describes how the antenna radiates electromagnetic energy in different directions. The 5G Wi-Fi antenna's radiation pattern is the region where the antenna emits the most energy.

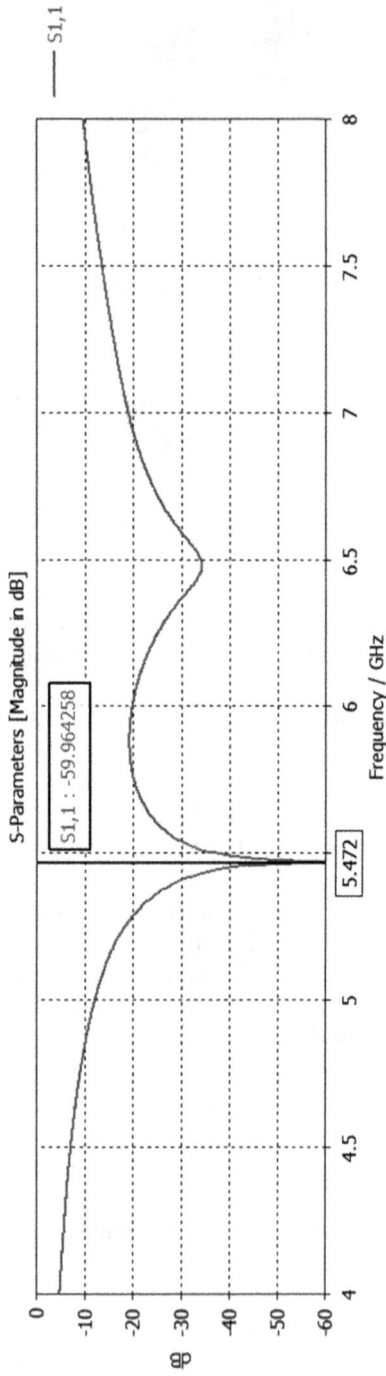

Figure 9.4 S11 or return-loss of Band-I.

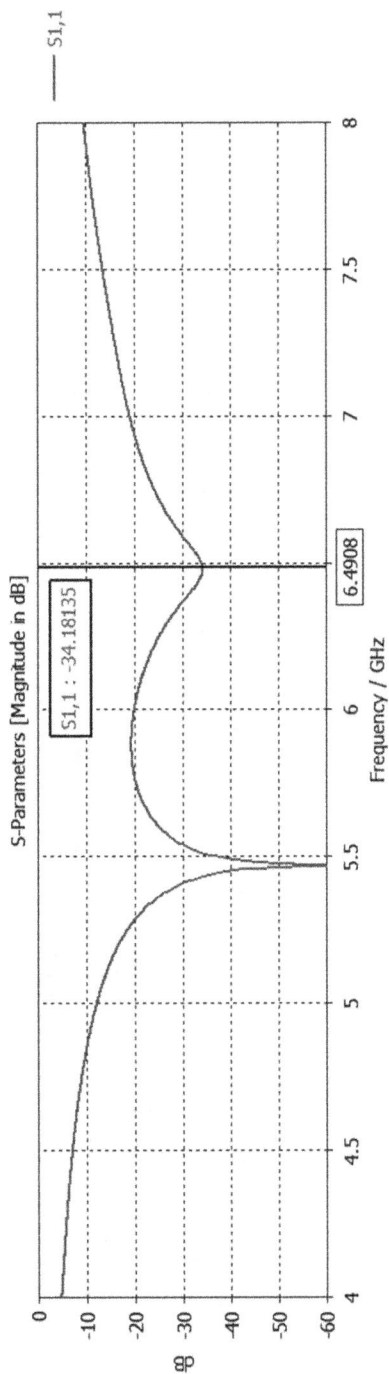

Figure 9.5 S11 or return-loss of Band-II.

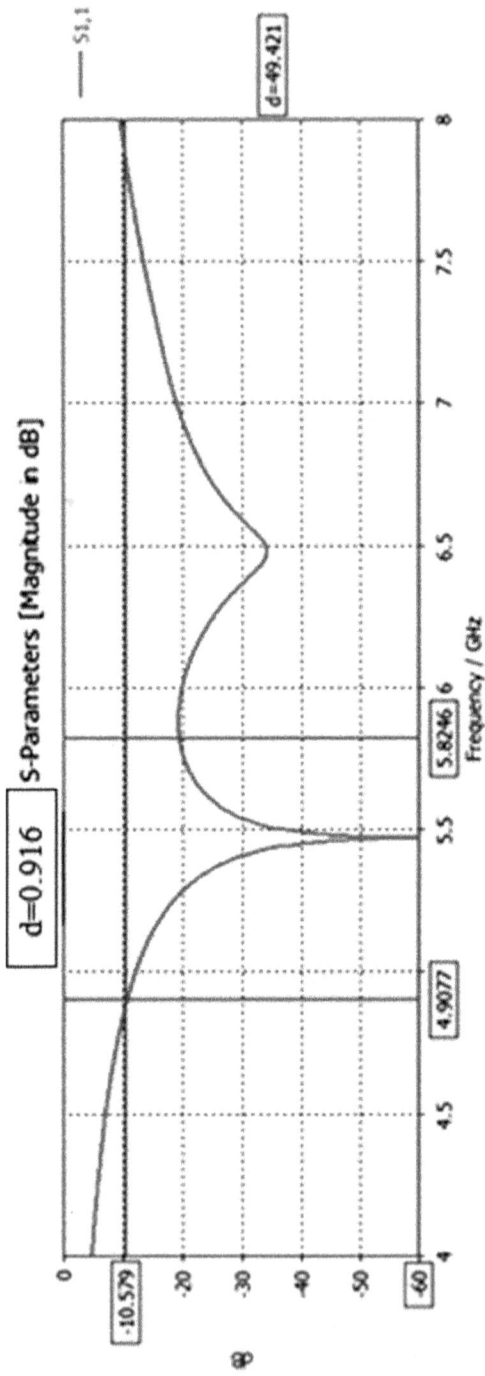

Figure 9.6 BW for Band-I.

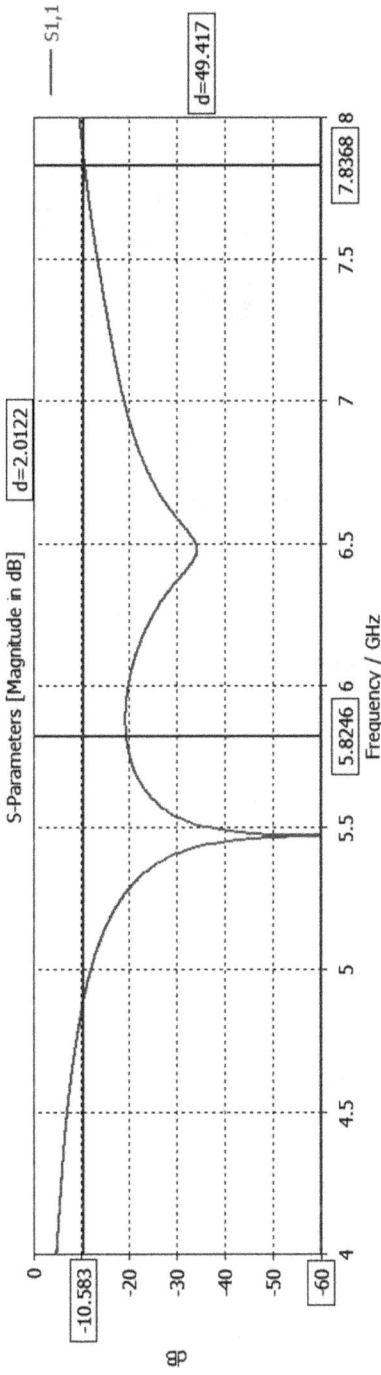

Figure 9.7 BW for Band-II.

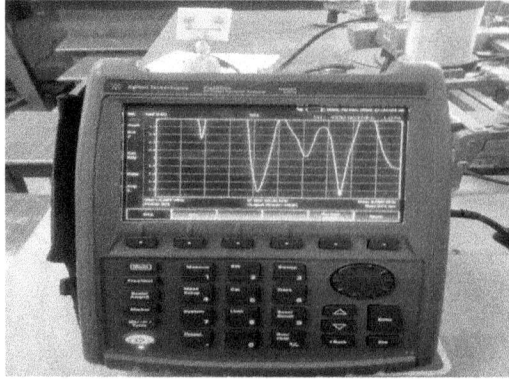

Figure 9.8 Antenna testing at VNA.

Table 9.2 Simulated and measured results of 1×2 antenna

	Simulated		Measured	
Name of parameter	Band-I	Band-II	Band-I	Band-II
Return-loss or S11	−59 dB	−34 dB	−50 dB	−28 dB
BW	961 MHz	2012 MHz	802 MHz	2000 MHz
VSWR	1.002	1.039	1.152	1.324
Resonant-frequency	5.4 GHz	6.4 GHz	5.4 GHz	6.35 GHz

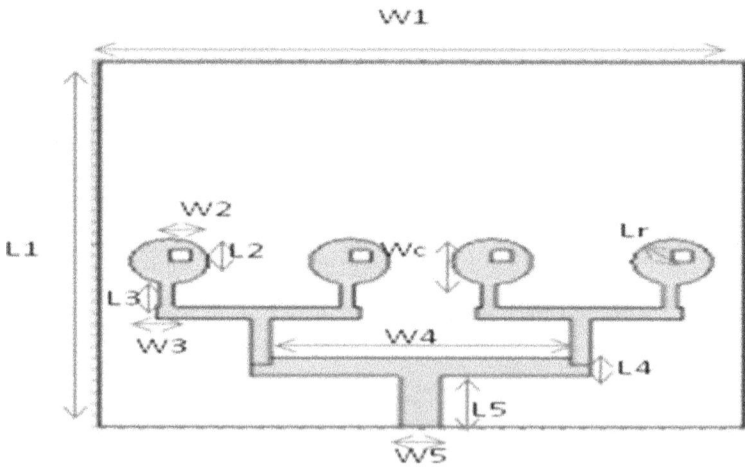

Figure 9.9   1×4 MPA design front view.

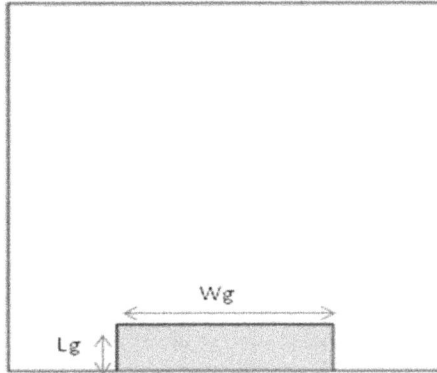

Figure 9.10 1×4 MPA design back view.

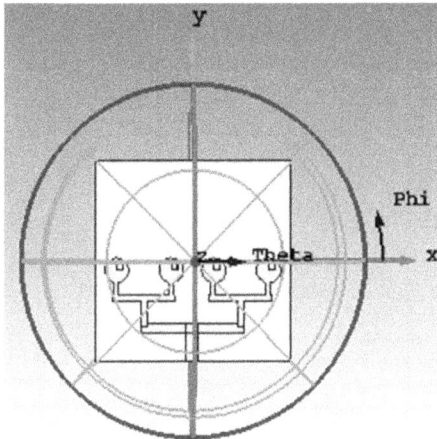

Figure 9.11 Antenna simulation in CST.

Figure 9.17 presents the surface current distributions of proposed designs. Wi-Fi antennas are essential components of wireless communication systems, and their surface current distributions directly influence radiation patterns, gain, impedance matching, and overall signal propagation. Simulation results of 1×4 antenna is shown in Table 9.4 and 9.5 shows the gain values.

Comparative results are shown in Table 9.6 and Optimized Resonant-frequency with 5G Wi-Fi channel is shown in Table 9.7.

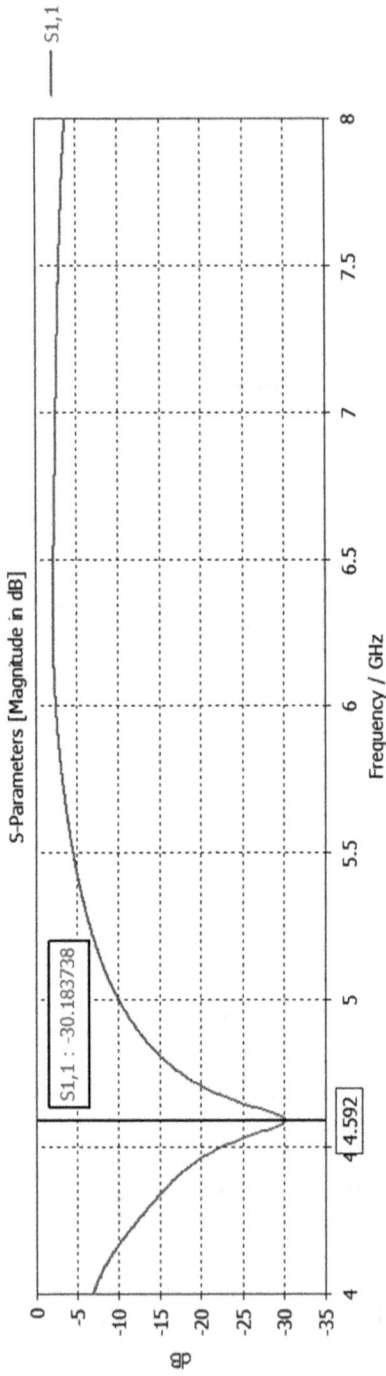

Figure 9.12 S11 or Return-loss.

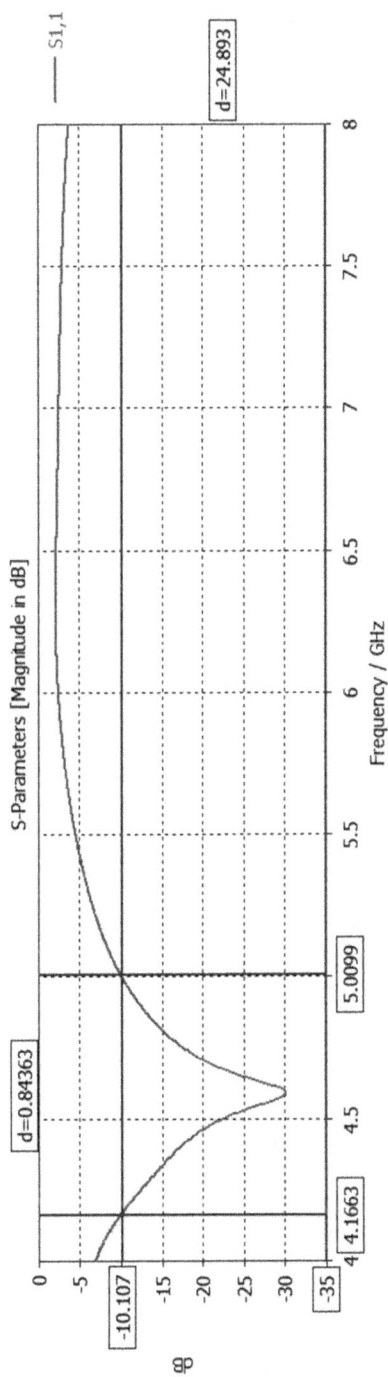

Figure 9.13 Bandwidth.

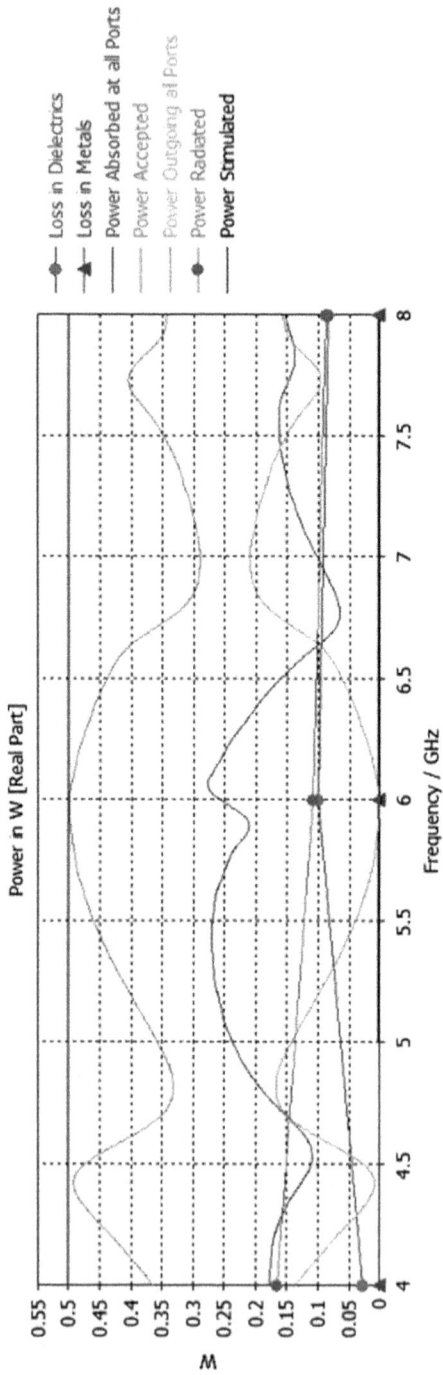

Figure 9.14 Antenna power.

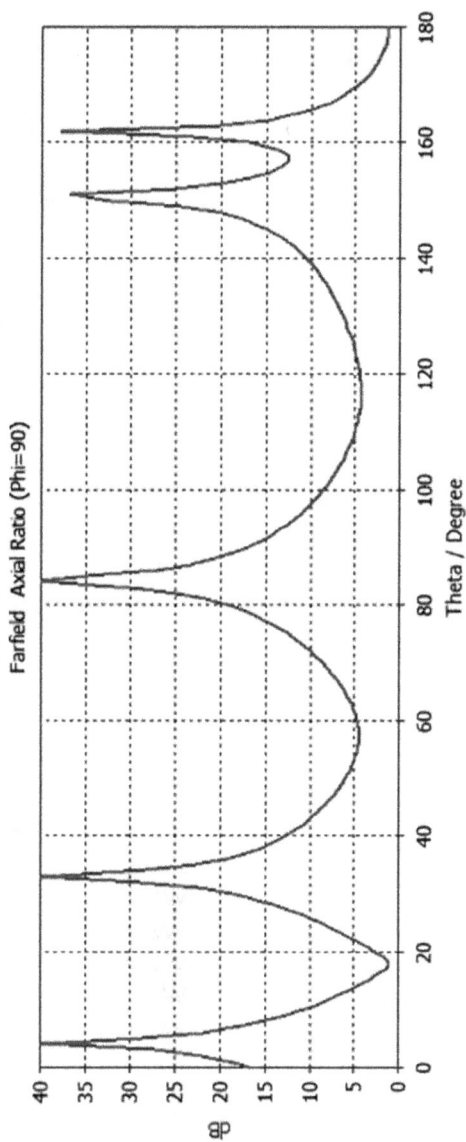

Figure 9.15 Polarization of antenna.

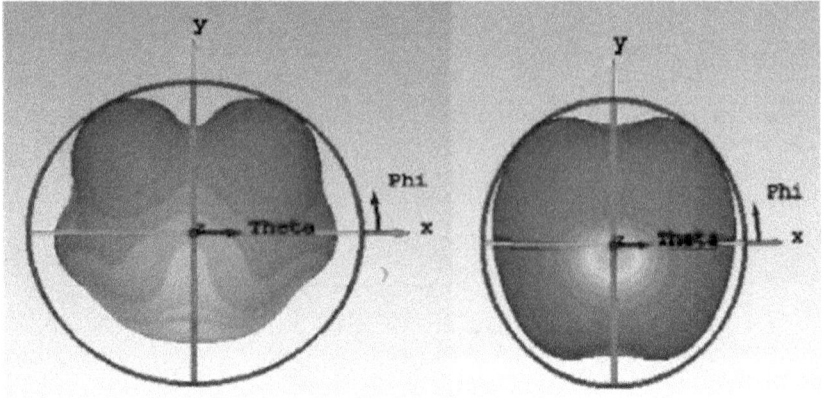

Figure 9.16 3D radiation pattern at 1×2 & 1×4.

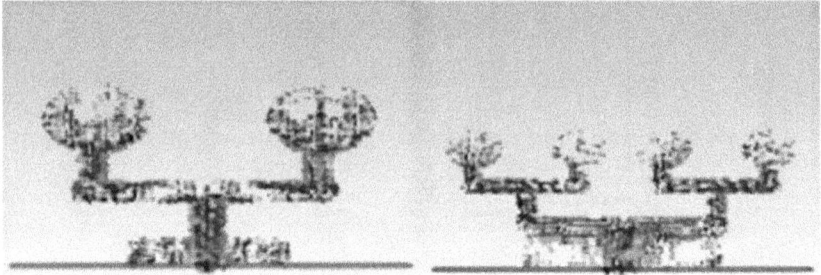

Figure 9.17 Surface current distributions at 1×2 & 1×4.

Table 9.3 The proposed antenna dimensions (in mm)

Length	Dimension	Width	Dimension
L1	32	W1	32
L2	1	W2	1
L3	2.06	W3	0.75
L4	1	W4	17
L5	6	W5	2
Lr	2	Wc	11.8
Lg	4	Wg	16

Table 9.4 Simulated results of 1×4 antenna

Sr no	Parameter	Simulated
1	S11 or Return$_{loss}$	−30 dB
2	BW	843 MHz
3	VSWR	1.063
4	Resonant$_{frequency}$	4.5 GHz

Table 9.5 Simulated value of gain

Array Antenna	1×2	1×4
Value of Gain	5.17 dBi	7.22 dBi

Table 9.6 Result comparison

Ref.	Dimensions (L × W) mm²	Frequency	Return-loss	Impedance BW
[4]	425 mm²	FR4	7.4 GHz	−17 dB
[14]	542 mm²	FR4	6.5 GHz	−22 dB
			5.4 GHz	−27 dB
Design-I [2×2]	314.16 mm²	FR4	5.4 GHz	−59 dB
			6.4 GHz	−34 dB
Design-II [1×4]	314.16 mm²	FR4	4.5 GHz	−30 dB

Table 9.7 Optimized resonant-frequency with 5G Wi-Fi channel

Sr no	Array type	Resonant-frequency	5G Wi-Fi channel
1	Design I [1×2]	5.4 GHz	UNII-2C
		6.4 GHz	UNII-5
2	Design II [1×4]	4.5 GHz	Licensed

9.8 CONCLUSION

A circular MPA-A can be an effective solution for a 5G Wi-Fi network due to its compact size and ease of integration. A circular MPA-A is designed and fabricated, which is suitable for new channels available under 5G communication. Optimized different resonant frequencies, 5.4 GHz, 6.4 GHz, and 4.5 GHz, with a maximum value of gain, is 7.22 dBi and BW 2012 MHz, are obtained. This antenna is excellent for applications requiring wireless communication via 5G Wi-Fi networks due to its compact size and dual frequency. Compared to the designs of other microstrip antenna arrays, the antenna's layout is simple and has a low weight and compact footprint.

The suggested antennas have an operational frequency that ranges between 4 GHz and 8 GHz, which covers numerous bands used for 5G communication and satisfies the fundamental BW requirements of Wi-Fi applications. The proposed designed antennas provide low return-loss, a wide BW, and improved gain characteristics.

REFERENCES

1. M. Alsukour, and Y. S. Faouri, "Side-Edge Dual-Band Antenna for 5G and Wi-Fi 6 Applications," *Microwave Mediterranean Symposium (MMS)*, Pizzo Calabro, Italy, pp. 1–4, 2022.
2. S. P. Pavithra, and K. Vidhya, "Design of a Circular Microstrip Patch Antenna with High Gain for 5G Applications using 5G Millimeter Wave Bands Comparing to Rectangular Microstrip Patch Antenna," *2nd International Conference on Technological Advancements in Computational Sciences (ICTACS)*, Tashkent, Uzbekistan, pp. 30–34, 2022.
3. S. Jabeen, Q. U. Khan, and S. Amin Sheikh, "A Compact Highly Isolated MIMO Antenna Design for WiFi-6E & 5G Applications," *19th International Bhurban Conference on Applied Sciences and Technology (IBCAST)*, Islamabad, Pakistan, pp. 973–978, 2022.
4. Y. M. Al Hyasat, Y. S. Faouri, D. I. Abualnadi, and M. K. Abdelazeez, "Two Element UWB Array Antenna for 5G Communications, WiFi-5 and WiFi-6 Applications," *2022 IEEE International Symposium on Antennas and Propagation and USNC-URSI Radio Science Meeting (AP-S/URSI)*, Denver, CO, USA, pp. 1048–1049, 2022.
5. Y. M. Hyasat, D. Abualnadi, and Y. S. Faouri, "Two Elements Ultra Wideband MIMO Antenna for 5G Communications, WiFi-5 and WiFi-6 Applications," *14th International Conference on Communications (COMM)*, Bucharest, Romania, pp. 1–4, 2022.
6. N. Praween, S. Singh, and S. Khan, "ReconFigure 9.5G sub 6 GHz and Wi-Fi bands Micro-strip Antenna using hybrid slotting and mix substrate material techniques," *IEEE International Conference on Advances and Developments in Electrical and Electronics Engineering (ICADEE)*, Coimbatore, India, pp. 1–8, 2020.
7. N. Singh and Y. Kumar, "Blockchain for 5G Healthcare Applications: Security and Privacy Solutions," *IET Digital Library*, pp. 1–30, 2021. DOI: 10.1049/PBHE035E
8. F. C. Gül, and S. Eker, "Microstrip Fed Circular Patch Antenna for 5G and Wireless Communication," *28th Signal Processing and Communications Applications Conference (SIU)*, Gaziantep, Turkey, pp. 1–4, 2020.
9. A. Assoumana, and G. Çakir, "Defected Ground Structure (DGS) based Miniaturization of a single Circular Patch Antenna for 5G Mobile Applications," *28th Signal Processing and Communications Applications Conference (SIU)*, Gaziantep, Turkey, pp. 1–4, 2021.
10. C. C. Chang, Y. F. Lin, M. -T. Nguyen, Y. X. Liu, H.-T. Chen, and H. M. Chen, "Design of MIMO Antennas for WiFi/5G Small Cell Applications," *International Symposium on Antennas and Propagation (ISAP)*, Taipei, Taiwan, pp. 1–2, 2021.

11. S. C. Das, L. C. Paul, M. N. Hossain, M. Z. Mahmud, and R. Azim, "A Dual Band Miniaturized Spiral-shaped Patch Antenna for 5G and WiFi-5/6 Applications," *IEEE International Conference on Signal Processing, Information, Communication & Systems (SPICSCON)*, Dhaka, Bangladesh, pp. 87–90, 2021.
12. R. A. Panda, A. Kumar Sharma, and M. Kumar, "Modified Array of Circular Patch Antenna for 5.9 GHz WLAN Application," *International Conference on Recent Innovations in Electrical, Electronics & Communication Engineering (ICRIEECE)*, Bhubaneswar, India, pp. 394–397, 2018.
13. N. P. Kulkarni, J. S. Kulkarni, N. B. Bahadure, and P. D. Patil, "Circular Patch Antenna with comb-shaped slot for NR-79/Wi-Fi Applications," *IEEE Region 10 Conference (Tencon)*, Osaka, Japan, pp. 905–908, 2020.
14. S. Khade, A. Chinchole, P. Pandey, S. Umredkar, V. Magare, and M. Sonkusale, "Circularly Polarized Cylindrical Slot Antenna With and Without DGS," *4th International Conference on Trends in Electronics and Informatics (ICOEI)*, Tirunelveli, India, pp. 309–313, 2020.
15. J. Colaco, and R. Lohani, "Design and Implementation of Microstrip Circular Patch Antenna for 5G Applications," *International Conference on Electrical, Communication, and Computer Engineering (ICECCE)*, Istanbul, Turkey, pp. 1–4, 2020.
16. S. R. Govindarajulu, R. Hokayem, and E. A. Alwan, "Dual-Band Antenna Array for 5.9 GHz DSRC and 28 GHz 5G Vehicle to Vehicle communication," *IEEE International Symposium on Antennas and Propagation and North American Radio Science Meeting*, Montreal, QC, Canada, pp. 1583–1584, 2020.
17. N. Cascalho, P. Pinho, V. Cruz, and N. B. Carvalho, "Dual-band printed circular monopole for Wi-Fi," *IEEE MTT-S International Microwave Workshop Series on 5G Hardware and System Technologies (IMWS-5G)*, Dublin, Ireland, pp. 1–3, 2020.
18. D. Sharma, V. N. Saxena, and P. Singh, "A simple plasmonic based rectangular microstrip patch antenna resonating at 5.9 GHz," *International Conference on Signal Processing and Communication (ICSC)*, Noida, India, pp. 386–389, 2015.
19. S. X. Ta, I. Park, and R. W. Ziolkowski, "Circularly Polarized Crossed Dipole on an HIS for 2.4/5.2/5.8-GHz WLAN Applications," *IEEE Antennas Wirel. Propag. Lett.*, vol. 12, pp. 1464–1467, 2013.
20. R. Tiwari, R. Sharma, and R. Dubey, "Design of 2×2 Microstrip Patch Antenna Array and Optimization of Bandwidth using Efficient Machine Learning Technique," *International Journal of Emerging Technology and Advanced Engineering*, vol. 12, no. 08, pp. 78–82, 2022.
21. R. Tiwari, R. Sharma, and R. Dubey, "Microstrip Patch Antenna Parameter Optimization Prediction Model using Machine Learning Techniques," *IJRITCC*, vol. 10, no. 9, pp. 53–59, 2022.

Chapter 10

Microstrip antenna for 5G wireless systems

Megha Gupta and Hemant Kumar Gupta
SAGE University, Indore, India

10.1 INTRODUCTION

In dealing with signals such as radio or electromagnetic (EM) waves, wireless communication is a method of transferring data between any numbers of devices that are not physically linked to one another. The range of these EM waves may only be a few meters [1]. This can be similar to Bluetooth technology or the millions of kilometersinvolved in space radio communication technologies. Thus, wireless systems cover a wide range of applications, including Bluetooth, infrared, satellite communication, wireless networks, mobile and radio communication, etc. [2]. These wireless communication technologies rely heavily on their antennas. The IEEE standard states that an antenna is a metallic structure that transmits and receives radio waves. With the least amount of complexity, a well-designed antenna enhances overall system performance [3].

10.2 OVERVIEW OF MICROSTRIP ANTENNA

Microstrip patch antennas (MPAs) are inexpensive, modest, lightweight, and easy to assemble. Because of these exceptional qualities, it is one of the most studied and used elements for several applications in wireless communication, mobile radio, radar engineering, military systems, space communication, and other areas. It has a radiating "patch" of metallization that is positioned above the grounded substrate and supplied with energy using the proper feeding mechanism [4]. The common patch excites the TM modes and is a perfect electric conductor (PEC).

10.2.1 Antenna parameters

The antenna's characteristics are evaluated using the far-field, close-range, and free-space ranges. Radiation intensity, gain, power flow, and polarization are among the radiation parameters of an antenna that are spatially represented. Such a pattern may be either directional or isotropic, effectively transmitting or receiving electromagnetic waves in one or both directions. This pattern, which is known as omnidirectional, may be both directed and non-directional in a certain plane (Figure 10.1).

DOI: 10.1201/9781003422440-10

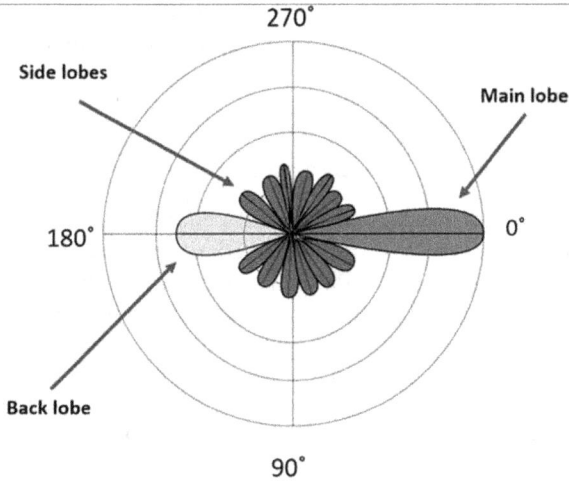

Figure 10.1 Omnidirectional radiation pattern of antenna.

10.2.2 Radiation intensity

The formula to calculate radiation intensity, or U (W unit solid angle), uses the power emitted by an antenna in a certain direction per unit solid angle.

$$U = r^2 W_{rad}$$

Where

$$(W/m^2)$$

It is the ratio of a radiation output in one direction to the average overall output of the antenna [5]. In the lack of a specific direction, the direction with the greatest intensity is taken into account.

10.2.3 Gain

It is the ratio of an isotopically radiating structure's intensity in one direction to another. This measurement takes the antenna's efficiency and directionality into consideration.

$$G = \eta e D$$

10.2.4 Antenna efficiency

The antenna's efficiency is the power differential between the antenna's emission and reception. This serves as an illustration of how antennas efficiently transmit output with a minimum of losses. Different factors, including as impedance mismatch, conduction, and dielectric losses, can result in losses.

$$\eta e = Prad/Pin$$

10.2.5 Voltage standing wave ratio

When there is an imbalance among the impedances of the antenna and the transmission line, waves that are reflected result in standing waves. The voltage standing wave ratio (VSWR) is calculated by dividing the standing wave's greatest voltage by its minimum value. A higher value of VSWR is implied by a greater mismatch.

10.3 ARBITRARY PATCH SHAPED MICROSTRIP ANTENNA

Wireless communication systems frequently use microstrip patch antennas (MPA). However, because of their tiny gain values and limited frequency spectrum, these antennas have significant practical drawbacks [6]. The range of frequencies where an antenna's properties adhere to a particular standard is thought of as the antenna's bandwidth. The impedance bandwidth is related to the input impedance, return loss, VSWR, radiation efficiency, etc [7]. A technique called ultrawideband (UWB) allows information to be transmitted over a wide bandwidth, and commercial, military, and medical uses are all calling for increasedcapacity [8]. For these wireless communication applications, what is required is UWB antennas that are small, affordable, resistant to jamming, and have a large bandwidth.

Wireless communication systems frequently use microstrip patch antennas (MPA). However, these antennas have two major operational disadvantages: narrow frequency bandwidth and small gain value [9].

10.4 MODIFIED CIRCULAR-SHAPED MICROSTRIP PATCH ANTENNA

Microstrip antenna includes selecting a design frequency, calculating the dimensions of the patch dependent on this frequency, selecting a method to feed for proper impedance matching, and modifying the shape of the patch to obtain the desired antenna characteristics. The frequency spectrum that is taken into account for design is the C-band (4 to 8 GHz), which has applications in the wireless telecommunications business, some WiFi devices, WLAN, weather RADAR systems, etc. The benefit of this frequency range is that it uses less bandwidth and performs better in bad weather. A microstrip emitting patch antenna with a circular form must have a 6 Ghz central frequency [5].

The side view of this MPA structure is shown in Figure 10.2.

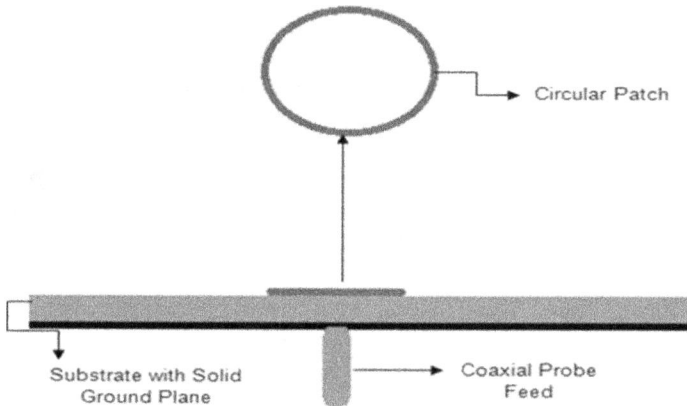

Figure 10.2 Radiant-circle side view of the MPA structure.

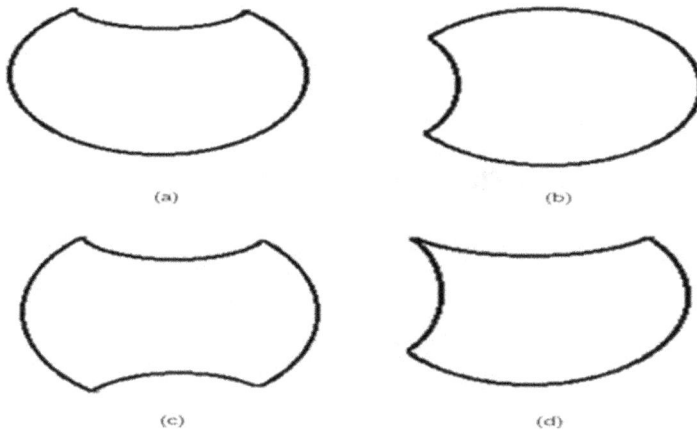

Figure 10.3 Different shapes of radiating patch obtained by modifying the circular shape with the removal of arc of length.

The patch's impedance and radiation characteristics are later improved. In order to adjust, an arc is removed from various spots on the metallic patch. The guided wavelength (g) inside the substrate cavity determines the length of the arc to be 14.4 mm (Figure 10.3).

10.5 INTRODUCTION TO MICROSTRIP PATCH ANTENNA WITH ARBITRARY SHAPE FIELD ANALYSIS

Employing numerical approaches that is the mixed potential integral equation (MPIE), finite-difference time-domain (FDTD), and the finite element method (FEM), the full-wave analysis of patch resonators with regular shapes is used to characterize the antenna properties. Numerical methods, including, among others, the MPIE, FDTD, FEMetc., are used in the full-wave analysis of patch resonators with regular shapes to describe the antenna properties.

The numerical analysis has been expanded to calculate numerous characteristics for MPAs of any shape, incorporating electric current distribution, input impedance, field patterns, and resonant frequencies. As was seen in the preceding section, altering the radiating patch's form enhances the antenna's impedance bandwidth, gain, and cross-polarization level. In order to generate an arbitrary form that can enhance antenna properties, a minor perturbation can be added to a conventional patch shape. The perturbation theory, in general, consists of a mathematical tool that approaches a complex system in terms of a simpler system having a known solution. You may get both perturbed and unperturbed electromagnetic wave solutions for the patch antenna's resultant E-field by using this perturbation method. The resultant field expression in curvilinear coordinates for a regular circular patch is validated using a soft HFSS simulator [10]. Then, using the suggested field analysis, two new, arbitrary-shaped patches in the C-band are created for the appropriate E-field patterns.

10.6 FIELD EVALUATION OF PERTURBED PATCH PROFILES

The antenna properties might be improved by this damaged construction. Therefore, this section creates a flexible tool that can be used to determine the E-field pattern of any random radiating patch [11].

In the plane of an orthogonal curvilinear coordinate system $(q1, q2, z)$, consider a damaged circular patch with an unusual shape [12]. The analytical function F of a complex variable $= x + jy$ [13] may be used to quickly convert a Cartesian coordinate system $(x, y,$ and $z)$ into a curvilinear coordinate system. In terms of real and imaginary components, this analytic function may be expressed as $F() = q = q1 + jq2$, where q1 acts as a constant patch boundary and $q2$ is perpendicular to it. Let G be an analytical function of the complex variable q in curvilinear coordinates, and let F be invertible to G. Any patch profile may be defined by this function $G(q)$ in a curvilinear coordinate system.

Let the curvilinear coordinate system's representation of the patch's circular represented in equation 10.1. In order to create patch $G1(q)$, the perturbation is now applied to the profile $G(q)$ under consideration. Patch $G1(q)$ is defined as:

$$G1(q) = eq\,(1 + (q)) \tag{10.1}$$

Where $G1(q)$ is used to scale the magnitude of the disturbance and $X(q)$ defines the perturbation.

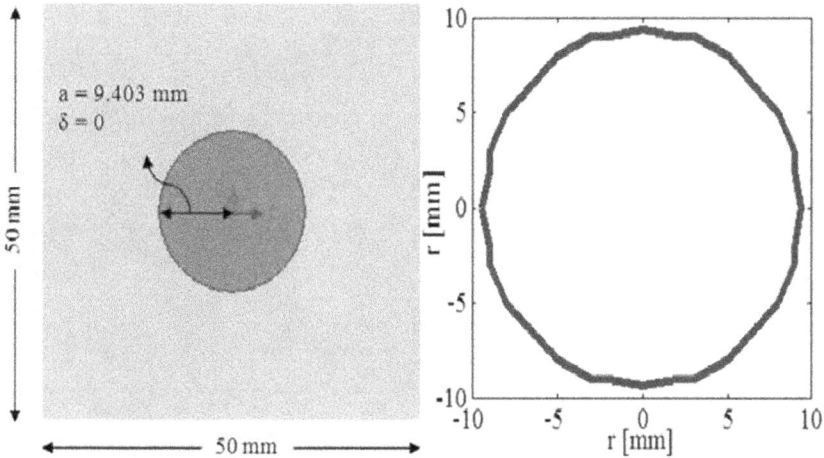

Figure 10.4 Circular shape (ϕ = 0°) (a) Layout of circular shaped MPA (b) Theoretical unperturbed circular patch shape.

10.7 CIRCULAR PATCH ANTENNA (ϕ = 0°)

Today, C-band is utilized for many different things, including as satellite communication, Wi-Fi devices, weather radar systems, etc... In order to radiate as much energy as possible at azimuth angle (xy plane) = 0°, a regular circular-shaped MPA is designed using this operating band.

At a design frequency of 6 GHz, the starting dimensions of a circular microprocessor route antenna are chosen using the conventional design methods. The rounded antenna patch is installed on top of the grounded dielectric substrate known as RT/Duroid. The patch has a 50 mm overall diameter and a relative permittivity of 2.2 (Figure 10.4).

10.8 ARBITRARY SHAPED ANTENNA (ϕ = −45°)

To alter the patch's assumed circular form and obtain the necessary radiation properties, the patch shape coefficients $G1(q)$ are adjusted [13]. As a result, a new random patch form is created by perturbing the circular patch's diameter, which has a maximum field strength directed at a 45-degree angle [14]. It displays the produced theoretical form $G1(q)$ = 0.029, while Figure 10.5 displays the corresponding arrangement of the suggested arbitrary shape (= 45°), maintaining the original circular MPA's structural configuration [15–20].

10.9 ARBITRARY SHAPED ANTENNA (ϕ = −180°)

The patch form $G1(q)$ coefficients are once again altered to get the maximum E-field strength along direction = 180°. The circular form is altered

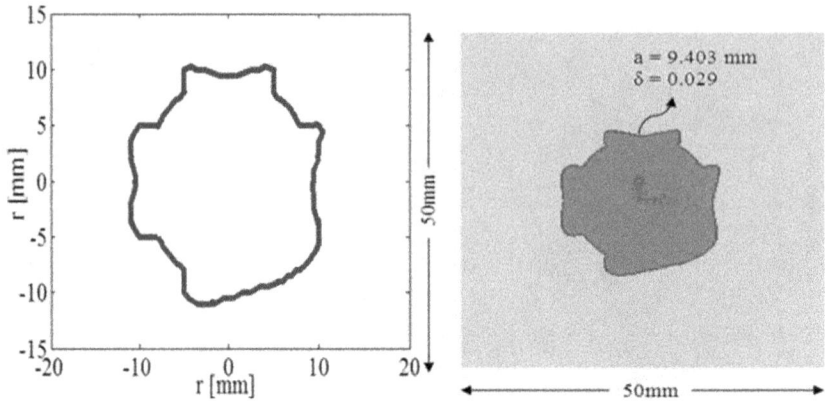

Figure 10.5 Arbitrary shape ($\phi = -45°$) (a) Theoretical arbitrary-shaped patch (b) Layout of arbitrary-shaped MPA.

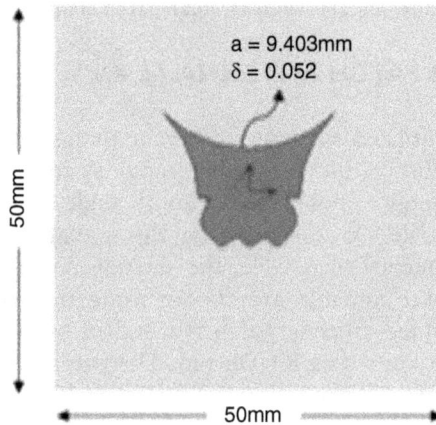

Figure 10.6 Arbitrary shape ($\phi = -180°$).

to take on a new arbitrary shape, as seen in Figure 10.6 with = 0.052. The figure maintains the original structural configuration of the circular MPA and also displays the matching architecture of the suggested arbitrary form (= 180°). This form is also used to verify the process, which produces measurable MPAs of any shapes for given field patterns [20–25].

The fabricated model is shown in Figure 10.6.

10.10 CONCLUSIONS

This chapterexhibits the effects of modifying geometry on the radiation and input impedance characteristics of a microstrip patch antenna in addition to showing the design analysis of alterations made to a radiating patch with a circular form. Gain, bandwidth, and cross-polar levels could all increase as the patch's form changes. The suggested shape is a circle with two eliminated arcs (from the top and left). 3-D EM Solver CST Studio Suite is used to model the structure. The suggested form has a broadside gain (= $0°$) of 7.95 dB, a cross-polar level of 85.37 dB, and a 200 MHz impedance bandwidth at 3 dB. Given its low cost and ease of production, this kind of antenna may prove to be of use in the military, business, and WiFi applications, among others.

The perturbation theory technique is also used to develop the E-field expression in order to construct the field analysis of an MPA with any shape. The necessary radiation pattern is produced, and a mathematically perturbed shape is specified in a curvilinear coordinate system in order to construct multiple probable patch configurations. The desired radiation pattern is established, and a mathematically perturbed shape is defined in a curvilinear coordinate system to create various possible patch shapes.

To validate the evolved field analysis, a standard circular MPA is initially created for a broadside radiation pattern (= $0°$). Later, two arbitrarily shaped patch antennas with maximum radiation in the directions of $45°$ and $180°$ are made using the results of this analysis. According to the output data, a circular polarization characteristic added to a designed MPA (= $45°$) at the resonance frequency increases its bandwidth to 230 MHz, and a different designed MPA (= $180°$) radiates in the exact opposite direction of the circular MPA while retaining its radiation properties.

REFERENCES

1. F. L. Lewis, "Wireless Sensor Networkss," *Smart Environments: Technologies, Protocols, and Applications*, John Wiley, New York, ISBN 0-471-54448-5, 2005.
2. A. Milligan, *Modern Antenna Design*, John Wiley & Sons Inc., IEEE Press, ISBN: 9780471720614, ebook, 2005 https://www.wiley.com/en-us/Modern+Antenna+Design%2C+2nd+Edition-p-9780471720614
3. D. Raychaudhuri, and N. B. Mandayam, "Frontiers of Wireless and Mobile Communications," *Proc. of IEEE*, vol. 100, no. 4, pp. 824–840, doi:10.1109/JPROC.2011.2182095, 2012.
4. C. A. Balanis, *Antenna Theory Analysis and Design*, J. Wiley & Sons Inc., New York, ISBN: 0-471-66782-X, 1997.
5. G. Addamo, O. A. Peverini, and R. Tascone, "A Ku-K Dual-Band Compact Circular Corrugated Horn for Satellite Communications," *IEEE Ants and Wireless Propag. Lett.*, vol. 8, 1418–1421, doi: 10.1109/LAWP.2010.2040573, 2009.
6. K. Yazdandoost, and R. Kohno, "Ultra-Wideband Antenna," *IEEE Communications*, vol. 42, no. 6, pp. 30–38, doi: 10.1109/MCOM.2004.1304230, 2004.
7. J. R. James, and P. S. Hall, *Handbook of Microstrip Antennas*, Peter Peregrinus Ltd., London, 1985.

8. N. Singh and Y. Kumar, "Blockchain for 5G Healthcare Applications: Security and Privacy Solutions," *IET Digital Library*, vol. 1, pp. 1–30, doi: 10.1049/PBHE035E, 2021.

9. K. Carver, and J. Mink, "Microstrip Antenna Technology," *IEEE Trans. on Ants. and Propag.*, vol. 29, no. 1, doi:10.1109/tap.1981.1142523, 1981.

10. J. R. Mosig, "Arbitrarily Shaped Microstrip Structures and Their Analysis With a Mixed Potential Integral Equation," *IEEE Trans. Microw. Theory Tech.*, vol. 36, no. 2, pp. 314–323, doi:10.1109/22.3520, 1988.

11. X. H. Yang, and L. Shafai, "Nodal-Based Basis Function for Full Wave Analysis of Microstrip Antennas with Arbitrary Geometries," *Electron. Lett.*, vol. 30, no. 11, pp. 830–831, doi:10.1049/el:19940608, 1994.

12. N. Singh, and G. K. Prajapati, "Wireless Sensor Network Security Problems-A Review", *International Journal of Emerging Technology and Advanced Engineering*, vol. 8, no. 1, pp. 77–80, 2018. https://www.ijetae.com/files/Volume8Issue1/IJETAE_0118_12.pdf

13. J. E. Ruyle, and J. T. Bernhard, "A Wideband Transmission Line Model for a Slot Antenna," *IEEE Trans. on Ant. and Propag.*, vol. 61, no. 3, 1407–1410, doi:10.1109/TAP.2012.2227100, 2013.

14. Y. T. Lo, D. Solomon and W. F. Richards, "Theory and Experiment on Microstrip Antennas," *IEEE Trans. on Ant. and Propag.*, vol. 27, no. 2, 137–145, doi:10.1109/TAP.1979.1142057, 1979.

15. T. Lo and S. W. Lee, *Antenna Handbook*, Van Nostrand Reinhold Inc., Springer Science, New York, ISBN: 978-1-4615-6459-1, 1988.

16. D. F. Moná Boada, and D. C. do Nascimento, "Including the Effects of Curvature in The Cavity Model and New Manufacturing Considerations for Cylindrical Microstrip Antennas," *IEEE Ant. and Wirel. Propag. Lett.*, vol. 17, no. 12, pp. 2419–2423, doi:10.1109/LAWP.2018.2877037, 2018.

17. W. K. Gwarek, "Analysis of Arbitrarily Shaped Two-Dimensional Microwave Circuits By Finite-Difference Time-Domain Method," *IEEE Trans. Micro. Theo. Tech.*, vol. 36, no. 4, 738–744, doi:10.1109/22.3579, 1988, 1988.

18. J. G. Maloney, G. S. Smith, and W. R. Scott, "Accurate Computation of the Radiation From Simple Antennas Using the Finite-Difference Time-Domain Method," *IEEE Trans. Ant. Prog.*, vol. 38, no. 7, pp. 1059–1068, doi:10.1109/8.55618, 1990.

19. J. G. Maloney, G. S. Smith, and W. R. Scott, "Accurate Computation of the Radiation from Simple Antennas Using the Finite-Difference Time-Domain Method," *IEEE Tran Ant. Prog.*, vol. 38, no. 7, pp. 1059–1068, doi:10.1109/8.55618.s, 1990.

20. P. A. Tirkas, and C. A. Balanis, "Finite-Difference Time-Domain Method for Antenna Radiation," *IEEE Trans. Ant. Prog.*, vol. 40, no. 4, pp. 334–340, 1992.

21. K. F. Lee, and K. F. Tong, "Microstrip Patch Antennas Basic Characteristics and Some Recent Advances," *in Proc. of IEEE*, vol. 100, no. 7, pp. 2169–2180, doi:10.1109/JPROC.2012.2183829, 2012.

22. Nitin I. Bhopale, and Sunil N. Pawar, "challenge for Designing a Multiband Ultra-Wide Band Antenna with its Solution," *IEEE Xplore*, doi:10.1109/CENTCON56610.2022.10051391, 03March, 2023.

23. Hui Dang, Lei Zhu, and Zhao An Ouyang, "Systematic Design Method for Mutual Coupling Reduction in Closely Spaced Patch Antennas," *Electronic*,

ISSN: 2637-6431. INSPEC Accession Number: 22919479, doi:10.1109/OJAP.2023.3262698, 2023.

24. H. Yang, X. Xi, and Y. Zhao, "Compact Slot Antenna With Enhanced Band-Edge Selectivity and Switchable Band-Notched Functions for UWB Applications," *IET Microw., Ants. & Propag.*, vol. 13, no. 7, pp. 982–990, 2019.

25. Quanhao Pang, and Gaoya Donga, "Metal Crack detection sensor based on Microstrip Antenna," *IEEE Sensors Journal*, vol.23no.8, pp. 122–131, 2023.

Chapter 11

Design and analysis of a high bandwidth patch antenna loaded with superstrate and double-L shaped parasitic components

Sateesh Kourav

Indian Institute of Information Technology, Design and Manufacturing, (IIITDM) Jabalpur, Jabalpur, India

Kirti Verma and Jagdeesh Kumar Ahirwar

Gyan Ganga Institute of Technology and Sciences Jabalpur, Jabalpur, India

M Sundararajan

Mizoram University, Aizawl, India

11.1 INTRODUCTION

11.1.1 A review of antennas

Voice communication is useful for communicating a notion, an idea, or even scepticism. The illustration below shows two individuals speaking. In this scenario, sound waves are employed to communicate. If two people separated by large distances want to communicate, sound vibrations must be translated into electromagnetic waves. An antenna is a device used to transform electromagnetic waves produced by data transmission [1]. Electrical impulses produce and emit electromagnetic waves using a portable and high-gain L-shaped microstrip patch antenna for 2.4 GHz industrial applications. Today, a wide variety of antennas are available, each with specific benefits and traits. You have the option of selecting the ideal antenna type for the task. The list of antenna types below should all be given some careful consideration. Antennas are attached to cables [2, 3]. It has certain advantages, such as mobility and cost, but it also has some drawbacks, such as restricted bandwidth. We propose the addition of alternate patch cuts, such as rectangular cuts and L shapes, to the radiating patch for Ku-band applications to boost the patch antenna's bandwidth. Patch slot antennas and printed antennas are the two types. Our studies are focused on microstrip patch antennas. A radiating patch of any planar shape is present on one side of a dielectric substrate, while the opposite side is covered by a ground plane.

DOI: 10.1201/9781003422440-11

In the resonance area, antennas are commonly utilized. The antenna is not tuned to the system's specified frequency range. Both the reception and the gearbox will suffer as a result. When a signal is broadcast to an antenna, it generates radiation that spreads in a predictable way over space [4, 5]. The radiation that is released into the atmosphere is represented by a radiation pattern. Microstrip antennas include lens antennas, reflector antennas, and array antenna.

11.1.2 The history of the antenna

Since Guglielmo Marconi created the first wireless communication system in 1895, one of the main goals of antenna research has been to discover ways to increase the amount of information bandwidth that an antenna can utilize while also maintaining high radiation efficiency and decreasing its physical size at any desired resonant frequency [2, 6, 7]. Compromises between these desirable qualities are usually needed when designing an antenna. In high-performance military applications, such as missiles, satellites, aircraft, and spacecraft, where there are restrictions on cost, weight, size, performance, and aerodynamic shape, low-profile antennas may be required [8]. Microstrip antennas satisfy these requirements, which is why they have increased in popularity in recent years. Around three decades ago, Deschamps, Glutton, and Bassinet proposed the notion of a microstrip antenna that could run simultaneously in the US and France. Lewin was quick to discover that radiation originates from strip line discontinuities. Kaloi continued his in-depth research into the fundamental rectangle and square geometries in the second half of the 1960s [1]. However, with the exception of the initial Deschamps work, no further research had been noted in the literature until Byron discovered a conducting strip line that was separated from a ground plane by a dielectric slab in the early 1970s. Along both radiating sides of this antenna, coaxial ports with a diameter of roughly half a wavelength and a length spanning several wavelengths were positioned at regular intervals. Soon afterward, Munson submitted a microstrip antenna patent application. When only one feed is used, an array feed's complexity, weight, and RF loss are decreased, A power-splitting network can be used to more effectively manage a number of orthogonal streams [9]. Additionally, there are several benefits to using circularly polarized microstrip antennas, including compactness, low weight, conformal mounting, compatibility with monolithic microwave integrated circuits, and compatibility with both millimeter wave and microwave integrated circuits [10]. This device has a high Q factor, low power and efficiency, polarization mixing, poor scanning performance, spurious feed radiation, and a very limited frequency bandwidth. A small portion of the total frequency under consideration, are just a few of the significant operational drawbacks of micro-strip antennas [11, 12]. It offers low efficiency due to dielectric losses and conductor losses. For some applications, such as secure communication systems, a narrow bandwidth is desirable. When a specific patch size and mode are selected, which will be small and suitable for usage on both curved and flat surfaces, it is both easier and less expensive to manufacture the antennas using existing printed-circuit technology; they are also better suited for hard surfaces. When applied, it is mechanically strong and compatible with MMIC designs. It's also extremely adaptable when it comes to operating frequency, polarization, radiation patterns, and input impedance. Similar standards are used in other public and commercial applications, such as wireless communications and mobile radio. In addition, interleaving loads like pins and varactor diodes between the patch and the ground plane enables the fabrication of adaptive components with programmable resonant frequency, impedance, bias, and pattern. Circularly polarized antennas with a single feed are becoming more common. Since circular polarization does not require the electric field vector to be aligned at the receive and transmit locations [13], it is preferable for current and future military and commercial uses. It is constrained in a number of obvious ways.

11.1.3 An antenna's objective

A device that sends and receives electromagnetic waves in empty space is called an antenna. Such a device connects a transmission line to a free area. It allows for the sending and receiving of signals [14]. The antenna serves as a channel for signal transmission. Through transmission cables, it sends a signal from the generator into open space. When an antenna is used to transmit or receive electromagnetic waves, it is referred to as an antenna. In the IEEE standard definitions of antenna terminology, according to the definition, an antenna is "a means of radiating or receiving radio waves," or, alternatively, between a void and the steering system.

11.1.4 Components of an antenna

Definitions of a number of criteria are required in order to define an antenna's performance. Below, we'll go over a few of the fundamental characteristics of antennas.

11.1.5 Media potency Haz anchoring and radiation

The performance and coverage of the system are affected by an antenna's properties. The way an antenna concentrates or directs the energy it receives or emits is determined by its radiation pattern. No antenna produces more radiation overall than the one that sends signals to its entry connection. The antenna's radiation patterns are often displayed as a polar graph for a

360-degree angular pattern scaled in terms of relative power (dB) on one of the two barrio plans. When the camp intensity exceeds 0.707% of the maximum voltage at the bulb's centre (measured in (V/m) [12, 15]. These points are noted by the medium power points. The relationship of the antenna's radiation pattern from forward to backward (f/b), as well as the side and rear light bulbs, are significant characteristics. In real-world applications, it is difficult to totally remove antenna side and rear lobes. The side and back lobes of the antenna may have a variety of effects on the performance of the antenna and system. Because the energy is coming from places other than the required zone of coverage, it is wasted rather than being sent to or absorbed by the side and rear lobes. The side and back lobes of a transmitter may be exposed to radiation that interferes with the operation of other receive systems. The radiation from various transmit locations may then be gathered by the side and back lobes of a receiver, which interferes with the system. The side and back lobes of the antenna may have a variety of effects on the performance of the antenna and system. First, because the energy is arriving from directions other than the targeted zone of coverage, it is squandered rather than being sent to or absorbed by the side and back lobes [16]. The operation of other receiver systems may be hampered by radiation from a transmitter's side and back lobes. If radiation from numerous transmit sites were to be caught by the side and back lobes of a receiver, the system may possibly become disturbed.

11.1.6 Polarization

The term "polarization" describes the spatial orientation of the electromagnetic wave's E-field (electric vector) component as it is being transmitted through the transmission system. Low-frequency antennas are often used due to ground effect-reflected waves and physical building approaches. Vertically polarized while high-frequency antennas are typically horizontally polarized. Little strip-shaped antenna microstrip antennas are one option for microwave and millimeter-wave applications.

11.1.7 A micro-strip antenna is used

Microstrip antennas are necessary for missile communication antennas. Microstrip radiators are used in tiny arrays by radar altimeters. A further illustration is the use of antennas in satellite and telephone communications. Satellite imaging systems have previously employed microstrip arrays. Between ships or buoys and satellites, patch antennas have been utilized for communication connections. Microstrip antennas are often employed in smart weapons due to their small size [15].

Pagers, the global positioning system (GPS), and the global system for mobile communication (GSM) are among the main uses for microstrip antennas. Due to its small size, low profile, and growing demand, the micro-strip antenna has gained appeal in recent years in both commercial and defence applications.

11.2 LITERATURE REVIEW

Zhang and Zheng, "Design of a double L-slot micro-strip patch antenna array for Imax/WLAN application using step width junction feed" [2] considers the Indian city of Melmaruvathur. The suggested antenna discussed in the article is 36.350 mm wide, 1.10 mm thick and 30.650 mm long. The suggested antenna is antenna-standard compliant and runs at 3.60 GHz to cover both the WLAN and the We MAX frequency bands. It has been tried in two different HFSS feed settings [17]. Before constructing the array, the performance of a single microstrip antenna is assessed. The resonant frequency is lower than the simulated resonant frequency. Meanwhile, the −10 dB bandwidth of the conventional microstrip circular patch antenna. In comparison to the 2.40 GHz frequency range, the calculated return loss values for the 3.50GHz frequency range are −22.0 dB higher.

Article [6] considers Dual-Band Dual-Polarized Slotted-Patch Antenna for In-Band Full-Duplex Applications," The authors of [18] offer a relative bandwidth of 30% from 11.60 GHz to 14.80 GHz, a footprint smaller than 13.130 mm × 8.860 mm is necessary. To minimize space while enhancing bandwidth, a polygonal design is advised. Using two of its widths, the polygon generates resonance at two distinct frequencies. To bring these diverse responses together, the middle of an antenna is pierced in the shape of a circle. Radom is composed of two Teflon-glass substrates that share the same characteristics as the preferred substrate; one is 1.58 mm thick, and the other is 2.500 relative permittivity and 0.020 tan.

11.3 DESIGN METHODOLOGY

An antenna for the Ku-band with various slits and notches was constructed and simulated using the HFSS 3D electromagnetic modelling application. The recommended microstrip antenna covers a wide frequency spectrum. The idea was realized using PTFE as the substrate and copper as the antenna material. The antennas given have a resonance frequency of 15 GHz, a return loss more than 50 dB, and a VSWR of less than one. There are several frequencies that are resonant at 12.20 GHz and 15.10 GHz, as well as at 11.160 GHz, 15.62 GHz, and 17.60 GHz, with return losses of 18.99, 23.026, and −18, and VSWRs of 0.79 and 0.59, respectively [19]. A Ku-band probe-fed microstrip patch antenna is used in this design. The antenna's dimensions are calculated using the accepted formula for rectangular patch antennas. The calculations above are used to create basic patch antennas: A good radiator must have a reasonable breadth and exceptional radiation efficiency.

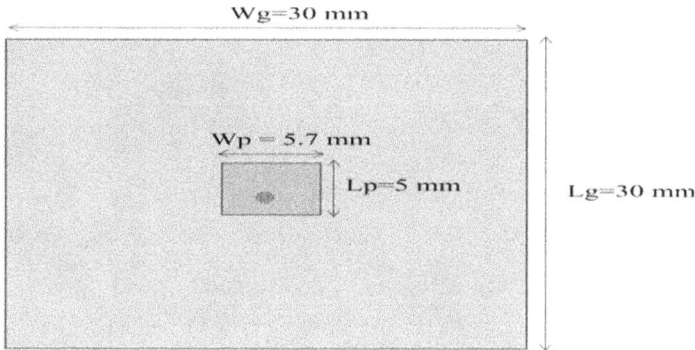

Figure 11.1 Antenna design.

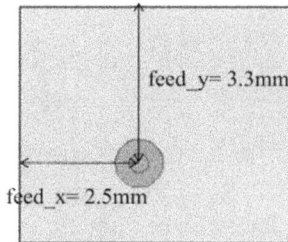

Figure 11.2 Feed locations.

11.3.1 Geometry of antennas (Design 1)

Using the fundamental principles described above, a straightforward microstrip antenna is produced using the Antenna Design Geometry of Antennas (Design 1), Figure 11.1 on the first try. This gadget operates as predicted in the Ku-band at frequencies between 12 and 18 GHz, a dielectric loss tangent of 0.0009 and a relative permittivity of 2.2.

A map of the earth in Figure 11.1 is depicted as $W_g = L_g = 30$ mm based on the foundation, and the patch measures $L = 5$ mm and $W = 5.7$ mm. The substrate is 1.6 mm tall and is made of Roger RT/duroid 5880. Simulation is employed to choose a feed point for the antenna rather than any empirical methods shown in Antennae Feed Locations in Figure 11.2. The feed point is at feed = 3.3 mm and feed = 2.5 mm in the patch's upper left corner. The best results in terms of impedance and current distribution come from this feed port configuration.

11.3.2 Geometry of antennas (Design 2)

The second phase involves the addition of two parasitic L-shaped pieces that span all four corners of the patch shown in Antenna Designs 2 (Figure 11.3). The result was an increase in the antenna's bandwidth and beamwidth.

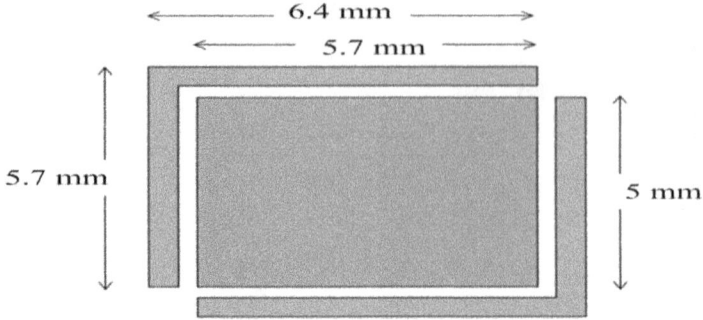

Figure 11.3 Antenna designs 2.

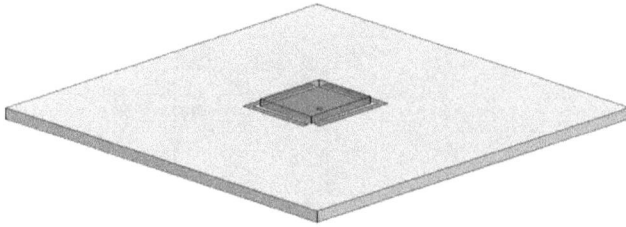

Figure 11.4 Antenna designs 4.

The L-shaped pieces are placed 0.2 mm apart from the primary patch. Electromagnetic induction may happen to any parasitic element that is even a minor wavelength distant from the host. The longer side of the L-section has two sides: one measuring 6.4 mm and the other, smaller side 5.7 mm. This L-strip has a 0.5 mm width.

11.3.3 Section 3 of antenna geometry (Design 3)

In the third iteration, the patch antenna's gain was increased by covering it with superstrata. The bandwidth was increased while the gain was decreased in the second iteration as a result of the addition of a parasitic component, as shown in Figure 11.4. A superstructure made of polystyrene with a dielectric constant of 2.6 is used. Although the superstrata are just 1 mm thick, its dimensions precisely match those of the active patch.

11.4 ANTENNA DESIGN 1 RESULTS

11.4.1 Return loss

Using a reference of −10 db, Antenna 1 has a bandwidth that varies from about 12 GHz to 18 GHz. The suggested innovation shows full Ku-band capabilities even in its most basic form shown in Figure 11.5.

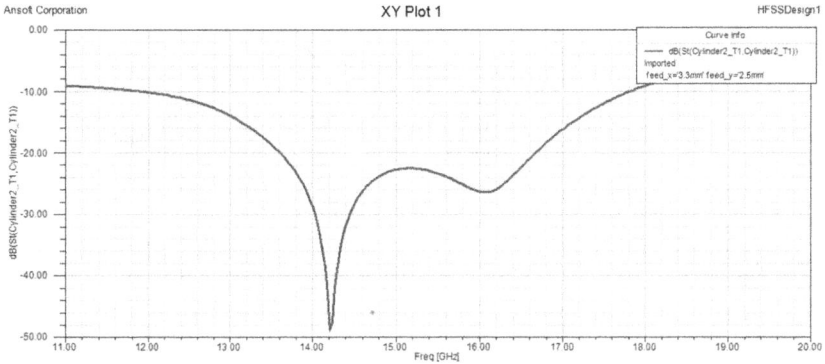

Figure 11.5 Return loss vs. frequency (Antenna design 1).

11.4.2 Radiation pattern

Broadside radiation is emitted by the antenna. In the x- and y-planes of Figure 11.6, the adiation pattern for design 1 is shown as red and black, respectively. The 3D radiation pattern in Figure 11.7 illustrates the complete anatomical structure of the lobe. Inconsistencies may be seen clearly in the x-y radiation pattern.

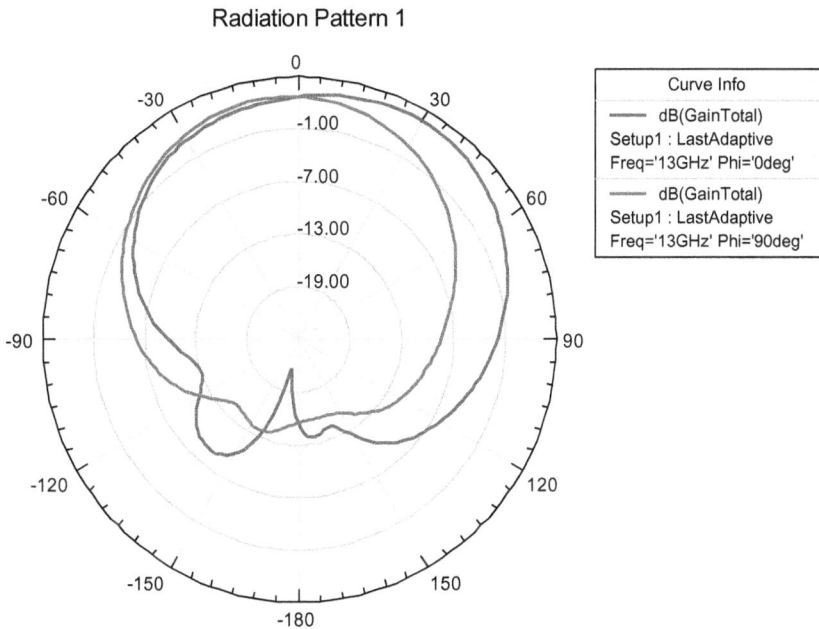

Figure 11.6 Radiation pattern for design 1.

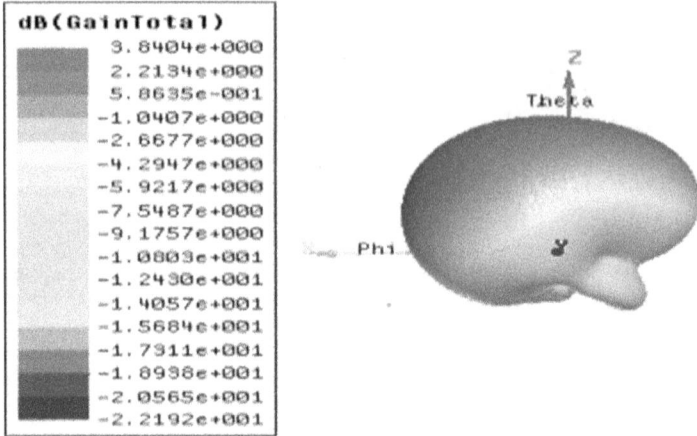

dB(GainTotal)
```
        3.8404e+000
        2.2134e+000
        5.8635e-001
       -1.0407e+000
       -2.6677e+000
       -4.2947e+000
       -5.9217e+000
       -7.5487e+000
       -9.1757e+000
       -1.0803e+001
       -1.2430e+001
       -1.4057e+001
       -1.5684e+001
       -1.7311e+001
       -1.8938e+001
       -2.0565e+001
       -2.2192e+001
```

Figure 11.7 First antenna design 3Dradiation pattern.

11.4.3 VSWR

The VSWR value is less than 2 and varies from 11 GHz to 16.4 GHz. The VSWR (Figure 11.8) appears to have the Ku-band in the wrong position.

11.4.4 Gain

The antenna's maximum gain is 4.24 dB. But there is a noticeable increase of more than 2 dB from 13.3 GHz to more than 18 dB, as shown in Figure 11.9.

11.5 RESULT OF ANTENNA DESIGN 2

11.5.1 Mention loss

There has been discovered to be a minimal loss of return at 14 GHz, which is −28, 4 dB. However, the return loss is less than −10 dB between 12 and 17.5 GHz in Figure 11.10.

11.5.2 Pattern of radiation

Since the initial iteration, the beamwidth of the radiation pattern has grown in both vertical directions. The obtained beamwidth is over 1200. The parasitic L forms on the microstrip antenna's sides are to blame for this. Figures 11.11 and 11.12 depict a 2D and a 3D radiation pattern, respectively.

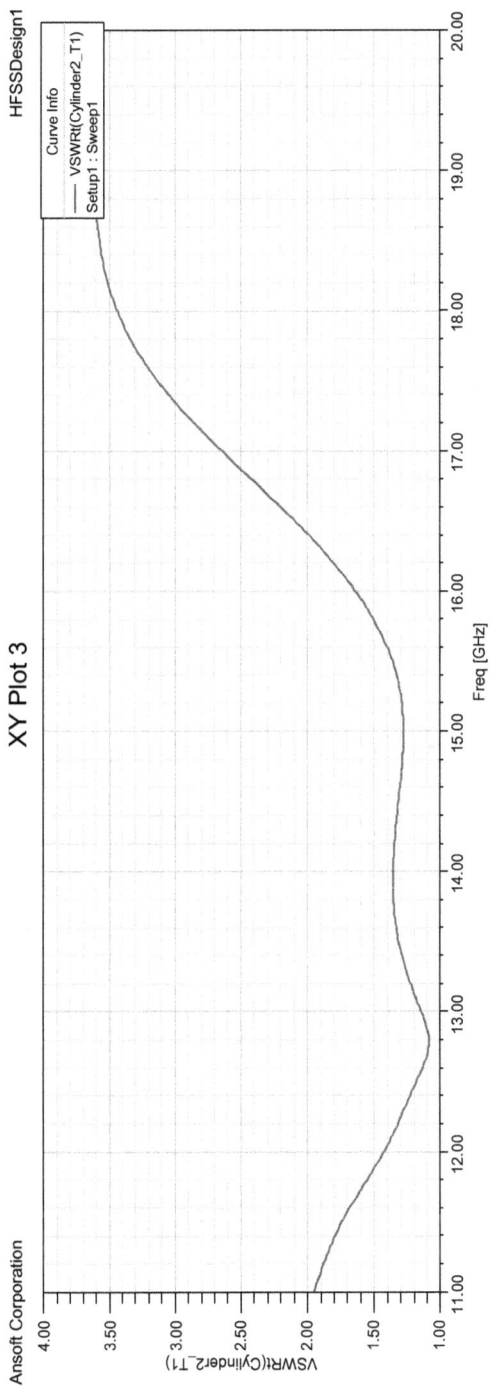

Figure 11.8 (Antenna design 1): VSWR vs. frequency.

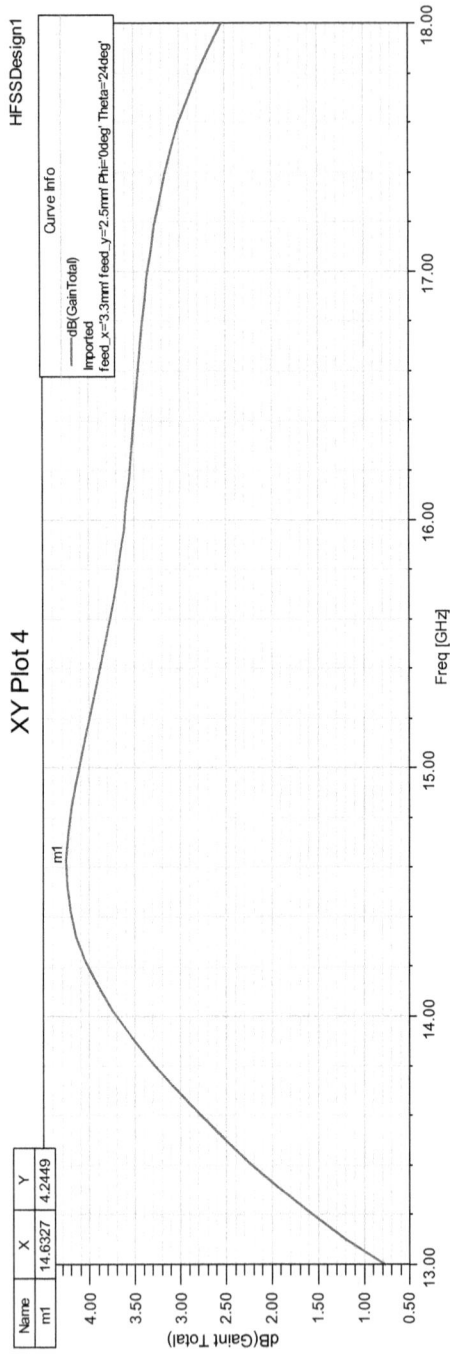

Figure 11.9 (Antenna design 1): Gain vs. frequency.

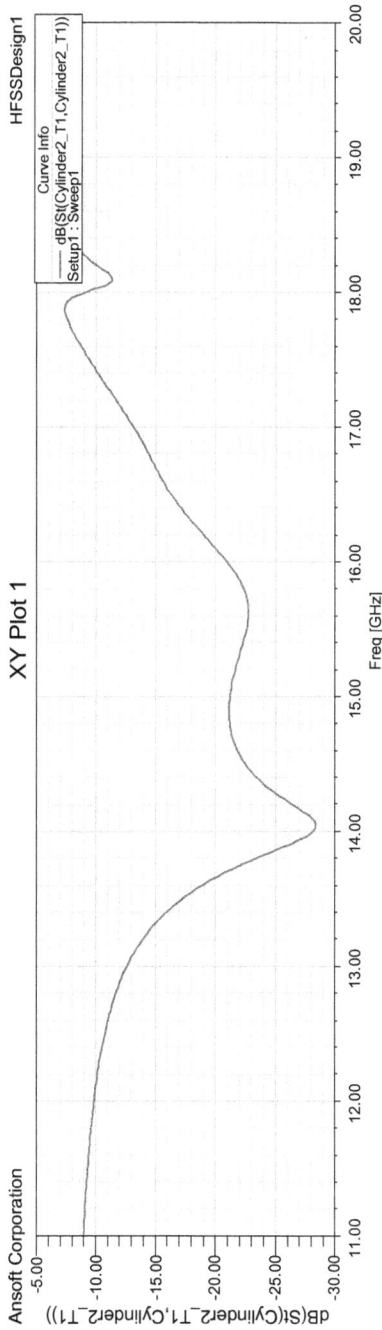

Figure 11.10 (Antenna design 2): Return loss vs. frequency.

Radiation Pattern 1

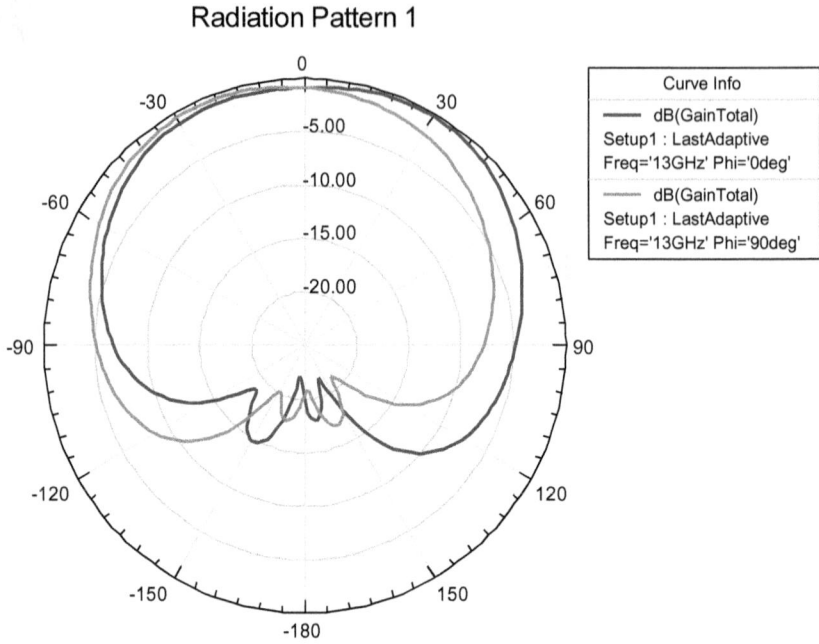

Curve Info
—— dB(GainTotal)
Setup1 : LastAdaptive
Freq='13GHz' Phi='0deg'
—— dB(GainTotal)
Setup1 : LastAdaptive
Freq='13GHz' Phi='90deg'

Figure 11.11 A second antenna design with a 2D radiation pattern.

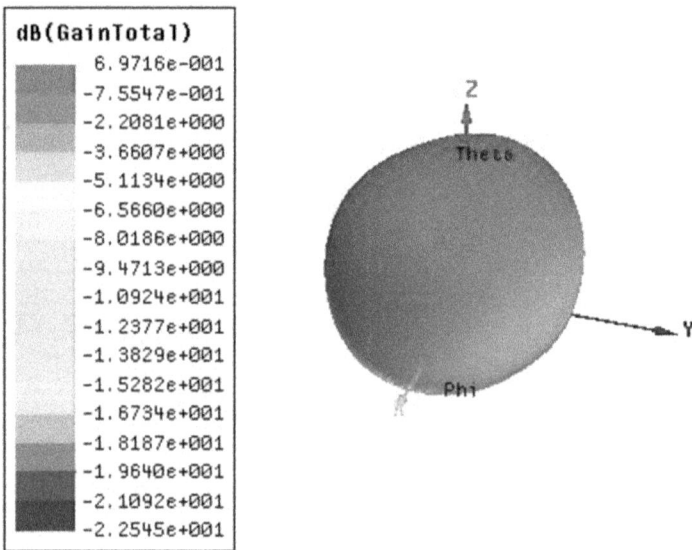

dB(GainTotal)

6.9716e-001
-7.5547e-001
-2.2081e+000
-3.6607e+000
-5.1134e+000
-6.5660e+000
-8.0186e+000
-9.4713e+000
-1.0924e+001
-1.2377e+001
-1.3829e+001
-1.5282e+001
-1.6734e+001
-1.8187e+001
-1.9640e+001
-2.1092e+001
-2.2545e+001

Figure 11.12 2D antenna design 3D radiation pattern.

11.5.3 VSWR

The VSWR bandwidth is further enhanced by VSWR Iteration 2 (Figure 11.13). VSWR is less than 2 between 11.7 GHz and 17.5 GHz.

The greatest gain is 4.3 dB at 14.24 GHz. The enlarged bandwidth includes the whole Ku-band, as seen in Figure 11.14.

11.6 ANTENNA DESIGN RESULTS 3

11.6.1 Mention loss

Over the whole Ku-band, the final antenna design model (Figure 11.15) provides the best return loss response. This can indicate that the bandwidth has grown.

11.6.2 Pattern of radiation

In comparison to the prior iteration, the addition of superstrata and parasitic L-shaped strips boosted both directivity and beamwidth. Figures 11.16 and 11.17 depict radiation patterns in 2D and 3D, respectively.

11.6.3 VSWR

The microstrip antenna design's third and final iteration offers the best VSWR response when compared to the other two in Figure 11.18. It also demonstrates that the bandwidth has been increased to encompass the entire Ku-band.

11.6.3.1 Gain

The greatest gain of the third design type is 4.68 at 14.57 GHz. The full gain response was transferred upstairs because the patch now has a super-state in Figure 11.19.

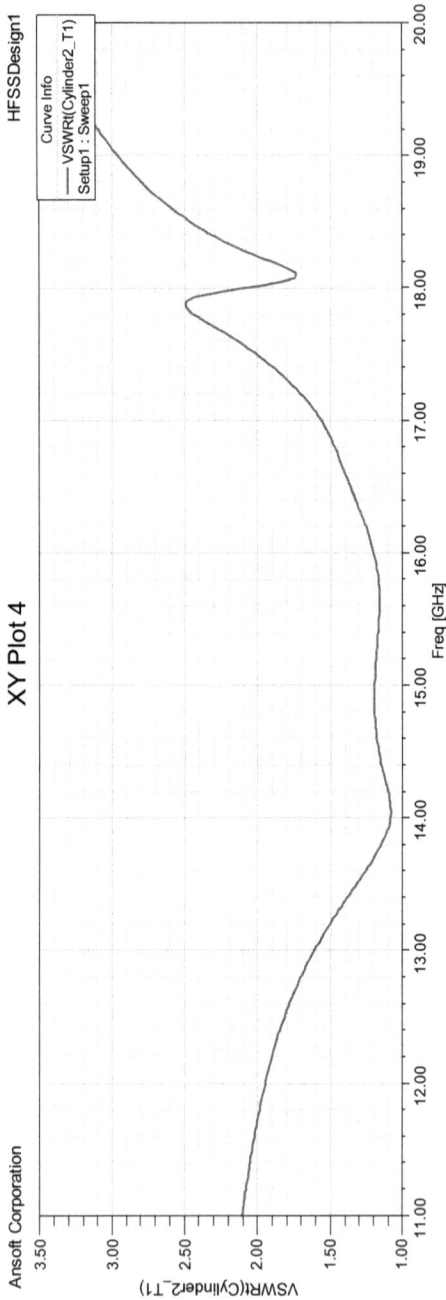

Figure 11.13 (Antenna design 2): VSWR vs. frequency.

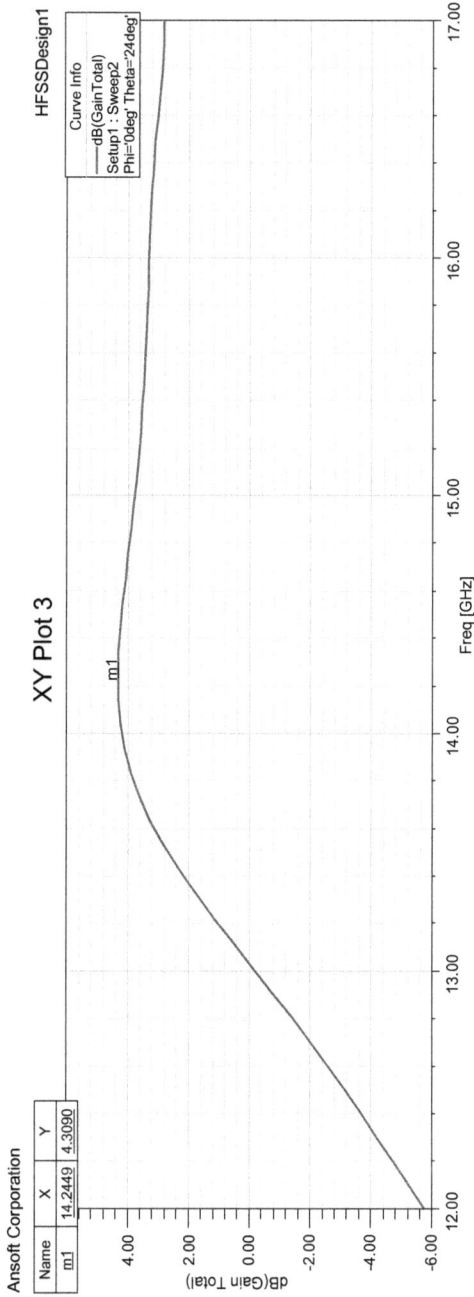

Figure 11.14 Gain Vs frequency (antenna design 2).

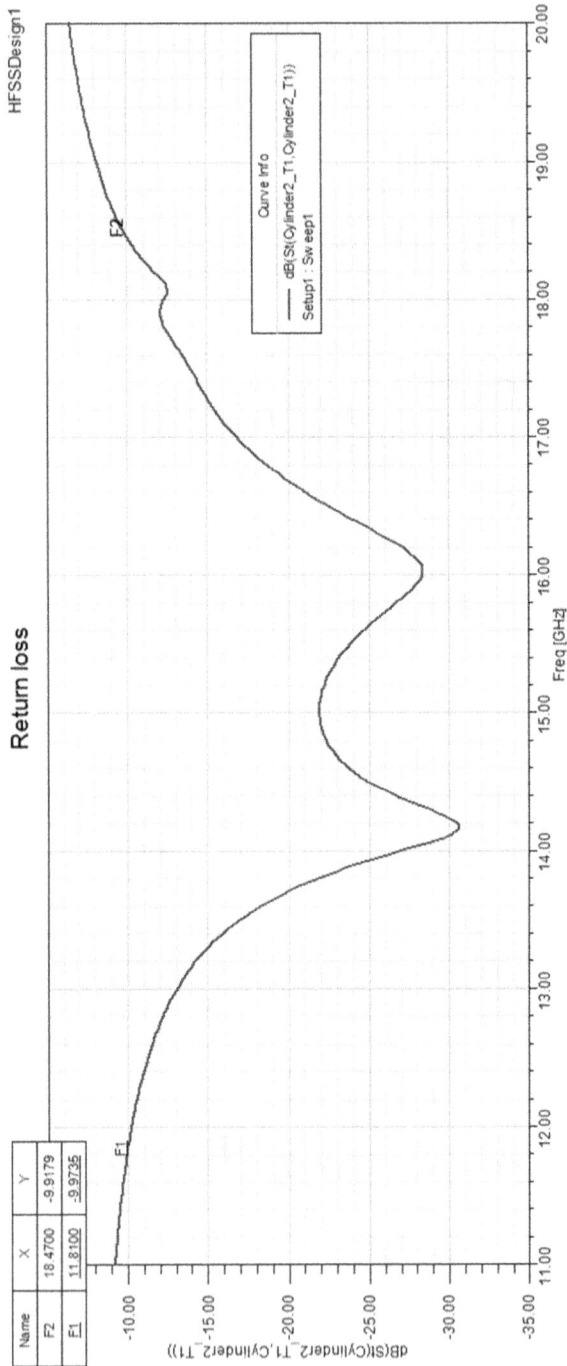

Figure 11.15 Return loss vs frequency (design of antenna 3).

Radiation Pattern 1

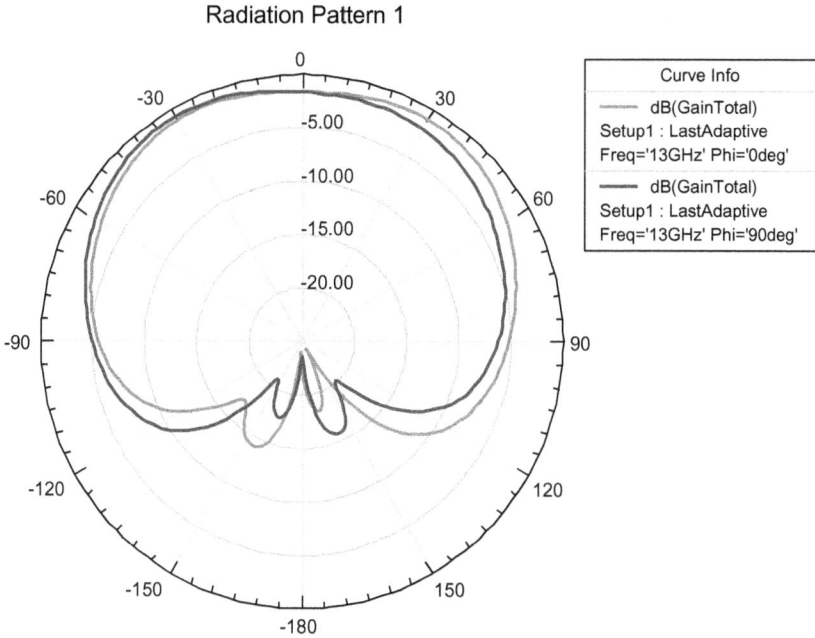

Figure 11.16 N-antenna design, a 2D radiation pattern.

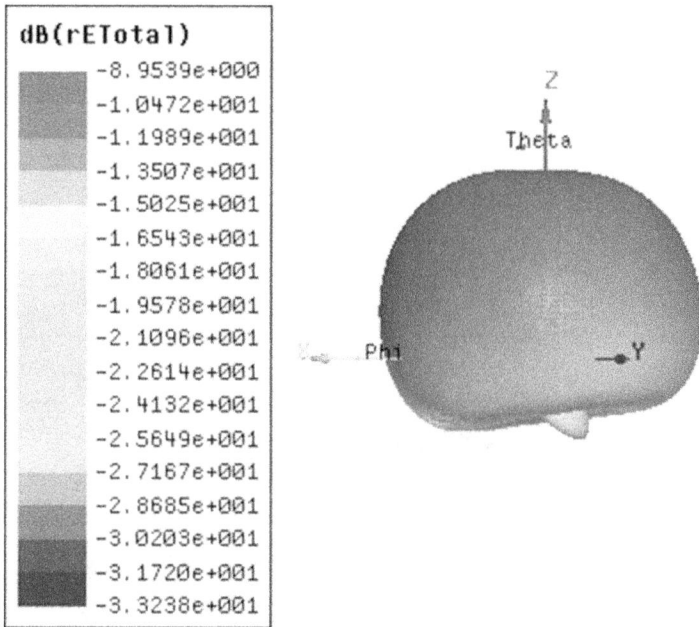

Figure 11.17 3D radiation pattern (design of antenna 3).

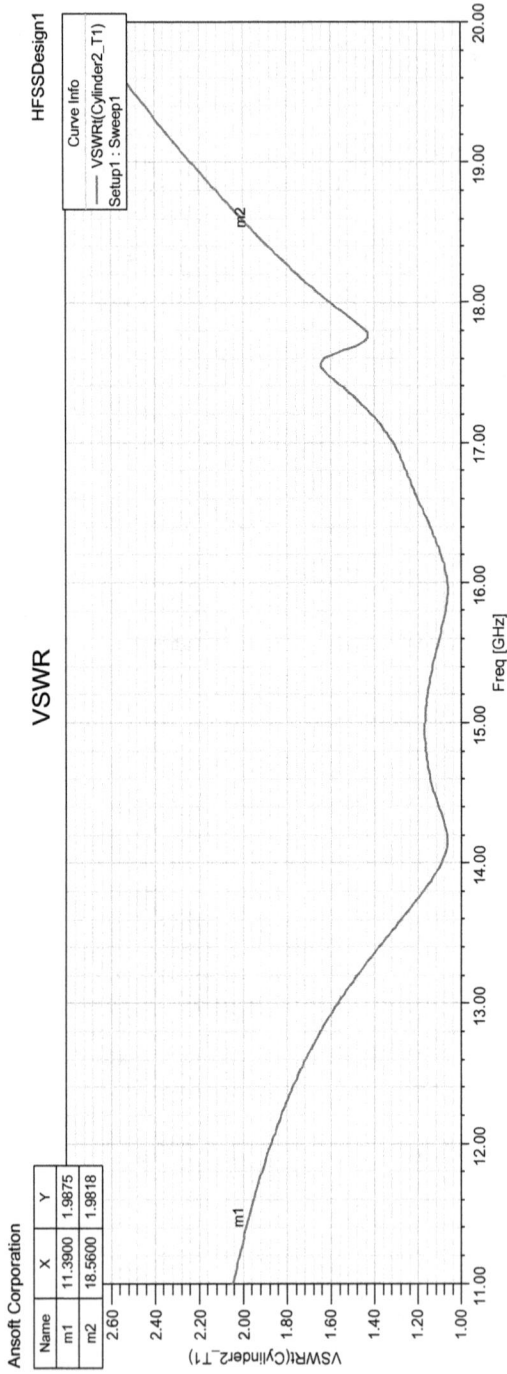

Figure 11.18 Compares frequency and VSWR (antenna design no. 3).

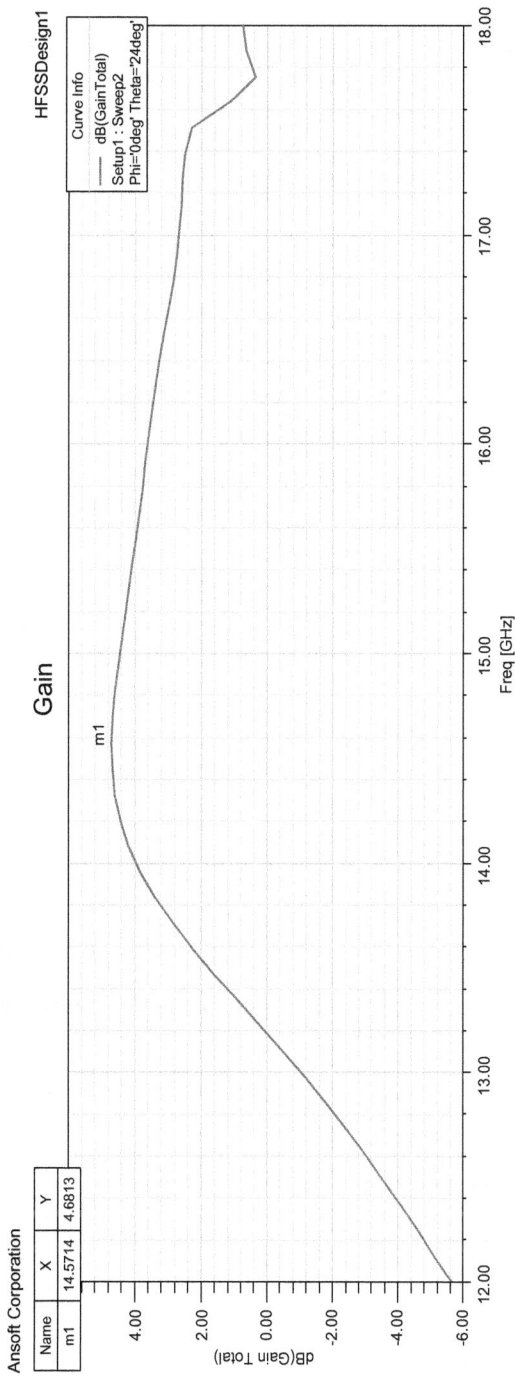

Figure 11.19 Gain versus frequency.

11.7 CONCLUSION

The suggested method develops an antenna that is supplied by a probe. It works throughout the entire Ku-band. An achievable bandwidth of 7 GHz exists. The operational frequencies are in the 12 to 18 GHz frequency region, and a VSWR of over 1.98 has been achieved. The parasitic L forms around the patch improve its beamwidth and bandwidth. Small Ku-band receivers can use the constructed antenna. Additionally, it can be used to fuel a parabolic dish that gets free satellite TV channels. Its small size guarantees that it won't interfere with the satellite signal's reception. In the current work, an antenna's gain is improved utilizing a superstrata that is quite simple to implement. More efficient superstrata with a frequency selective structure can boost gain and minimize unnecessary harmonics. To boost the antenna's gain, a superstratum is put above the patch. The results of the simulation reveal that the antenna operates brilliantly.

REFERENCES

1. Ashish K. Adiga, and Debdeep Sarkar, "CSRR and Stub Loaded Miniaturised Tri-Band Patch Antenna for 5G Base Station Application," *2022 IEEE Microwaves, Antennas, and Propagation Conference (MAPCON)*, pp. 1651–1656, 2022.
2. Tianhu Zhang, and Qi Zheng, "Ultrabroadband RCS Reduction and Gain Enhancement of Patch Antennas by Phase Gradient Metasurfaces," *IEEE Antennas Wirel. Propag. Lett.*, vol. 22, no. 3, pp. 665–669, 2023.
3. Inzamam Ahmad, and Ernesto Limiti," Design and Analysis of a Photonic Crystal Based Planar Antenna for THz Applications," *Electronics*, vol. 10, pp. 10–16, 2021.
4. Adelaida Heiman, "Design of a Conventional Horn Antenna for High Frequency Band," *2020 International Workshop on Antenna Technology (iWAT)*, 25–28 Feb. 2020.
5. Biao Hao, and Tong Li, "A Coding Metasurface Antenna Array with Low Radar Cross Section," *Acta Physica Sinica*, vol. 69, no. 24, pp. 244101, 2020.
6. Tran Hien Bui, and Nghia Nguyen-Trong, "Dual-Band Dual-Polarized Slotted-Patch Antenna for In-Band Full-Duplex Applications," *IEEE Antennas Wirel. Propag. Lett.*, vol. 22, no. 6, pp. 1286–1290, 2023.
7. Tao Zhang, and Xiaoyi Wang, "C-Band Linear Polarization Metasurface Converter with Arbitrary Polarization Rotation Angle Based on Notched Circular Patches," *Crystals*, vol. 12, no. 11, p. 1646, 2022.
8. Neng-Wu, and Guang Fu," Frequency-Ratio Reduction of a Low Profile Dual Circular Polarized Patch Antenna Under Triple–Resonance," *IEEE Antennas Wirel. Propag. Lett.*, vol. 19, no. 10, pp. 1689–1693, Oct. 2020.
9. Zhechen Zhang, and Shiwen Yang, "In-Band Scattering Control of Ultra-Wideband Tightly Coupled Dipole Arrays Based on Polarization-Selective Metamaterial Absorber," *IEEE Trans. Antennas Propag.*, vol. 68, no. 12, pp. 7927–7936, 2020.

10. N. Singh and Y. Kumar, "Blockchain for 5G Healthcare Applications: Security and privacy solutions," *IET Digital Library*, vol. 1, pp. 1–30, 2021. DOI:10.1049/PBHE035E
11. Jinhyeok Park, and Songcheol Hong, "Wideband Bidirectional Variable Gain Amplifier for 5G Communication," *IEEE Microw. Wirel. Compon. Lett.*, vol. 33, no. 6, pp. 691–694, 2023.
12. Ritu Vyas, and Suman Sharma, "Bridged Concentric Circular Microstrip Patch Antenna for C, X and High Frequency Band Applications," *Mater. Today: Proc.*, vol. 74, no. 2, pp. 392–400, 2022.j
13. Xiaosong Liu, Zehong Yan, Enlin Wang, Xiaofei Zhao, Tianling Zhang, and Fangfang Fan, "Dual-Band Orthogonally-Polarized Dual-Beam Reflect-Transmit-Array With a Linearly Polarized Feeder," *IEEE Trans. Antennas Propag.*, vol. 70, no. 9, pp. 8596–8601, 2022.
14. Youngmin Kim, Hongjong Park, and Sangmin Yoo, "High-Efficiency 28-/39-GHz Hybrid Transceiver Utilizing Si CMOS and GaAs HEMT for 5G NR Millimeter-Wave Mobile Applications," *IEEE J. Solid-State Circuits*, vol. 6, pp. 1–4, 2023.
15. Amit Birwal, "Design and Analysis of Compact Triple band Microstrip Patch Antenna," *2022 2nd International Conference on Emerging Frontiers in Electrical and Electronic Technologies (ICEFEET)*, pp. 1–6, 2022.
16. Mohammed Amine Bouchra Madouri, and Bachir Abes, "Novel Design and Optimization of S Band Patch Antenna for Space Application by Using a Gravitational Search Algorithm," *International Journal on Interactive Design and Manufacturing (IJIDeM)*, pp. 1131–1148, 2022.
17. Lin Tian, and Xin Zhou, "An Ultra-Wideband Microstrip Slot Antenna based on Metasurface," *2022 21st International Symposium on Communications and Information Technologies (ISCIT)*, pp. 135–138, 2022.
18. Amin Kianinejad, and Anxue Zhang, "Compact Dual-Band High-Efficiency Antennas Based on Spoof Surface Plasmon Polaritons," *IEEE Transactions on Antennas and Propagation*, vol. 71, no. 1, pp. 1075–1080, 2023.
19. A. Arroyo, and M. Casaletti, "Linear Polarization from Scalar Modulated Metasurfaces," *2022 16th European Conference on Antennas and Propagation (EuCAP)*, pp. 1–4, 2022.

Chapter 12

Bandwidth enhancement of microstrip patch antenna using metamaterials

Shravan Kumar Sable
LNCT, Bhopal, India

Puran Gour
NIIST, Bhopal, India

12.1 INTRODUCTION

Over recent years, metamaterial plays an important role in the implementation of novel microwave components and antennas because of their unusual properties. There are mainly two types of approaches which have been proposed to realize metamaterial transmission lines: 1. the L-C loaded approach, for the planer transmission line, which consists of shunt inductance and non-resonant lumped (chip or printed) series capacitance used as loading components [1, 2]. The design and realization of miniaturized antennas is one of the attractive applications of the transmission line based metamaterial [3–5]. The artificial electric and magnetic characteristics of metamaterials can be exploited to enhance radiation properties of electronic or photonic circuits, including antennas [6]. It has been proven that metamaterial can be applied to overcome limits of shrinking the antenna size [7]. The microstrip patch antennas have been one of the most popular antennas in current wireless communication systems as they have several desirable attributes, such as a low-profile planar configuration, being lightweight, having a simple design principle, low fabrication costs, and so on. Recently, a great deal of attention has been given to the properties of the artificial materials, named metamaterials, due to their interesting anomalous electromagnetic features and the wide variety of applications.

12.2 DESIGNING MICROSTRIPANTENNAS USING METAMATERIAL

The use of metamaterial to designany microstrip antenna basically consists of two parts:

1. Mathematical Analysis: Initially, we have to identify the substrate of the antenna for designing microstrip antenna by metamaterials; for this purpose, various parameters must be analyzed by mathematical

DOI: 10.1201/9781003422440-12

formulae and numerical analysis etc. There are many feeding techniques available, of which microstrip feeding techniques is adopted in this work.

2. Antenna Design using IE3D software: For the simulation of proposed antenna we have used IE3D software version 9.0. Different steps are followed for the design of a microstrip antenna using metamaterial antenna discussed in [8].

12.2.1 Mathematical analysis

There are three essential parameters that should be known while performing mathematical analysis [9, 10]:

Frequency of operation (f_0 = 2.5GHz)
Dielectric constant of the substrate (ε_r = 4.4)
Height of dielectric substrate (h = 1.6 mm)

Formula for calculation of microstrip line width and length
Designing of the microstrip line:
The width "w" of the microstrip line can be calculated as [9]

$$\frac{w}{h} = \begin{cases} \dfrac{8e^A}{e^{2A}-2} & \text{for } \dfrac{w}{h} < 2 \\[2em] \dfrac{2}{\pi}\left[\begin{aligned} B-1-\ln(2B-1)+ \\ \dfrac{\varepsilon_r-1}{2\varepsilon_r}\left\{\ln(B-1)+0.39-\dfrac{0.61}{\varepsilon_r}\right\} \end{aligned}\right] & \text{for } \dfrac{w}{h} > 2 \end{cases}$$

Where

$$A = \frac{Z_0}{60}\sqrt{\frac{\varepsilon_{r+1}}{2}+\frac{\varepsilon_r-1}{\varepsilon_r+1}\left(0.23+\frac{0.11}{\varepsilon_r}\right)}$$

$$B = \frac{377\pi}{2Z_o\sqrt{\varepsilon_r}}$$

where "Z_0" is the characteristic impedance of the microstrip line and is taken as 50Ω for the design.

The length "L_{ml}" of the microstrip line can be expressed as

$$L_{ml} = \frac{\lambda_{eff}}{4}$$

Where

$$\lambda_{\text{eff}} = \frac{c}{f\sqrt{\varepsilon_{\text{eff}}}}$$

The "w" and length "L_{ml}" of the microstrip feed line are calculated. Formulae used for mathematical calculations:

$$W = \frac{c}{2f\sqrt{(\varepsilon_r + 1)/2}} \qquad (12.1)$$

$$\varepsilon_{\text{reff}} = \left(\frac{\varepsilon_r + 1}{2}\right) + \left(\frac{\varepsilon_r - 1}{2}\right)\left[1 + 12\frac{h}{W}\right]^{-1/2} \qquad (12.2)$$

$$\Delta L = 0.412h\frac{(\varepsilon_{\text{reff}} + 0.3)\left(\dfrac{W}{h} + 0.264\right)}{(\varepsilon_{\text{reff}} + 0.258)\left(\dfrac{W}{h} + 0.8\right)} \qquad (12.3)$$

$$L = \frac{c}{2f\sqrt{\varepsilon_{\text{reff}}}} - 2\Delta L \qquad (12.4)$$

where:
  f = Operating frequency
  ε_r = Permittivity of the dielectric
  $\varepsilon_{\text{reff}}$ = Effective permittivity of the dielectric
  W = Patch's width
  L = Patch's length
  h = Thickness of the dielectric

12.2.2 Designing of the metamaterial structure

The metamaterial structure is proposed on ground plane as shown in Figure 12.1, the microstrip feed line which is designed on the exciting side of substrate. The exciting slot is placed at the centre of ground plane at a height of 1.6 mm. The length L_{ml} of the microstrip feed line, which does not include the ground plane, behaves as a stub matching for the slot. By varying the length of the matching stub, the reflection coefficient of the antenna can be controlled and this improves the return loss level accordingly. The internal gap of the meander slot is elected as 1.5 mm. Variation in the internal height of the slot tends to lead to variation in the return loss level and impedance bandwidth.

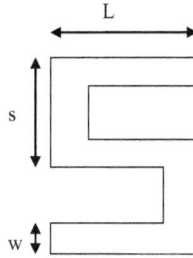

Figure 12.1 Geometry of proposed antenna (l = 10.25 mm, s = 7.5 mm and w = 1.5 mm.).

12.3 ANTENNA DESIGN BY USING IE3D SOFTWARE

The proposed antenna is designedusing FR-4 substrate. Different parameters, such as return loss, gain, directivity and the radiation pattern, can be obtained by using the EM Simulator IE3D software. Based on the simulations and mathematical computation we get the length and width of microstrip patch of microstrip antenna, which is calculated as shown in Table 12.1.

To reduce the antenna size and enhance bandwidth we use metamaterial structure on defected ground surface. By using proper slot cutting or stub matching in between the exciting patch and ground planes, the maximum power is transferred at edge of the cutting slot due to impedance matching. Designing parameters forproposedantenna are shown in Table 12.1.

12.4 SIMULATION AND RESULTS

The final patch of proposed antenna is shown in Figure 12.2 which is simulated by IE3D software, and Figure 12.3 shows the geometry of the proposed antenna.

The geometry consists of the, ground plane, which is S-shaped cut and microstrip line as a patch at height of 1.6 mm. Figure 12.4 shows the 3D view of the proposed antenna.

Table 12.1 Parameters usedfor the design of antenna

S.no	Parameters	Values for proposed antenna
1	Design frequency (f_0)	2.5 GHz
2	Dielectric constant (ε_r)	4.4
3	Height of substrate (h)	1.6 mm
4	Loss tangent of FR4	0.019
5	Width of microstrip patch (W)	3.05513 mm
6	Length of microstrippatch (L)	16.4478 mm

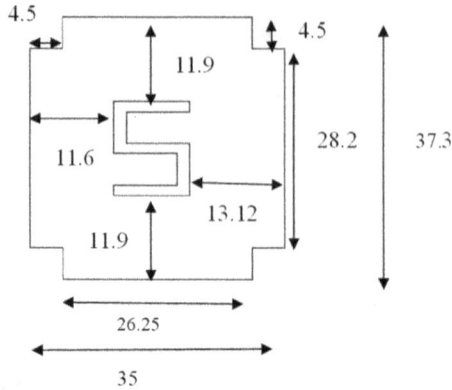

Figure 12.2 Proposed ground structure.

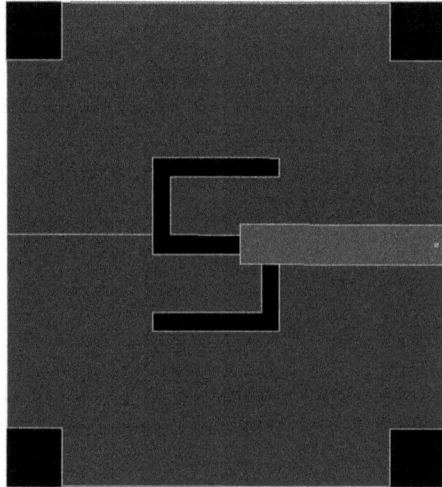

Figure 12.3 Geometry of proposed antenna (Top view).

The obtained minimum value of the antenna return loss is −27 dB at a frequency of 2.5 GHz as shown in Figure 12.5. Thus, the bandwidth obtained from the return loss result is 73.36%.

Moreover, the proposed antenna VSWR is less than 2 for 2.4–4.1 GHz and 5.1–6.3 GHz as shown in Figure 12.6. While VSWR indicates how much the antenna matches with that of cable impedance. There may be less reflection back from the source if the value of VSWR is less. The obtained value by our antenna poses a good value due to a lower mismatch; however, ahigh VSWR indicates that the port is on the upper side of the mismatch. Figure 12.7 shows the curve between directivity vs frequency for proposed antenna

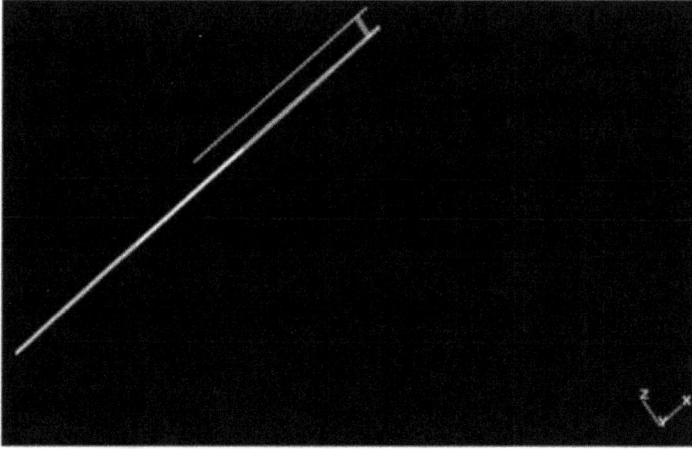

Figure 12.4 3D view of the proposed antenna.

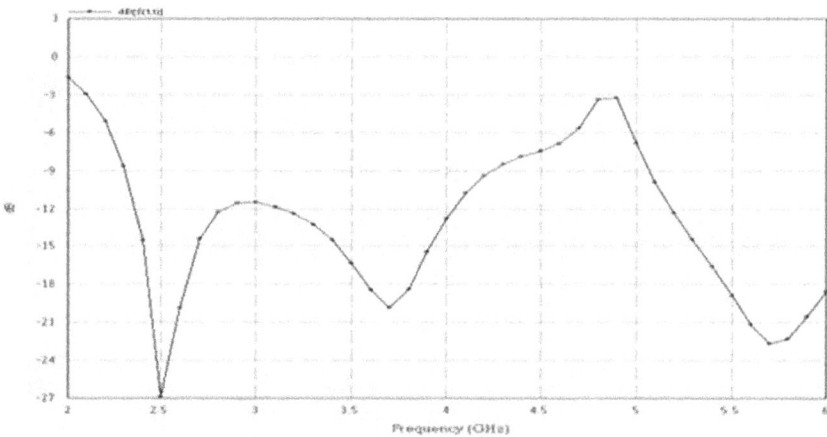

Figure 12.5 Return loss for the proposed antenna.

The maximum directivity of proposed antenna is 6.2 dBi.

The maximum gain of proposed antenna is 4.2 dBi, as shown in Figure 12.8. Figure 12.9 shows a Smith chart for the proposed antenna.

The 3D radiation pattern of the antenna is shown in Figure 12.10. The given radiation pattern indicates stable radiation pattern for the entire operating band.

The 2D radiation pattern in the polar plot shows the elevation pattern of the proposed antenna which has a maximum gain of 4.2 dBi as shown in Figure 12.11.

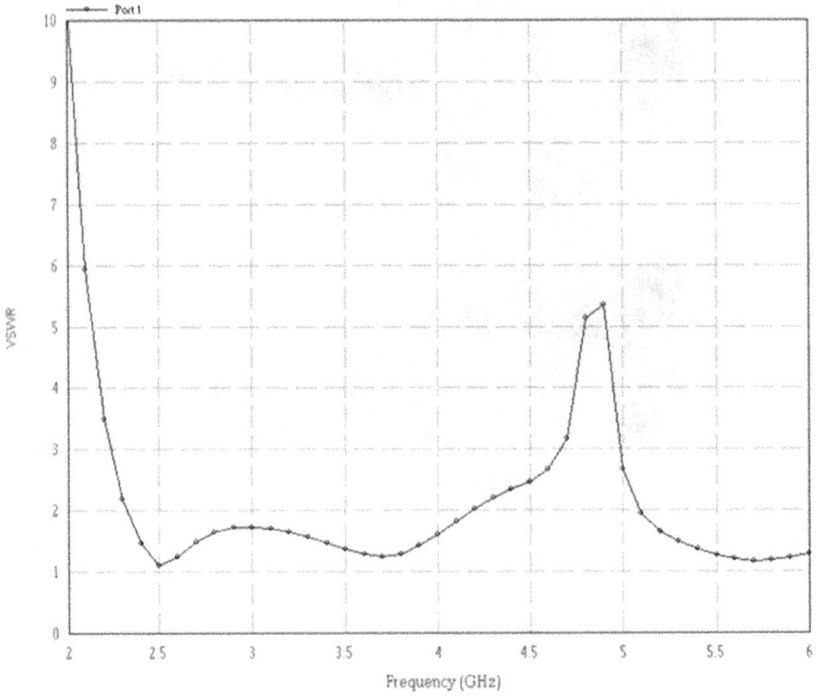

Figure 12.6 VSWR curve for proposed antenna.

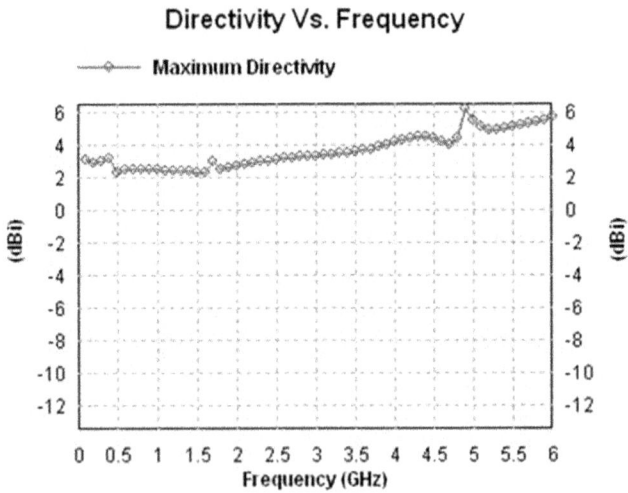

Figure 12.7 Directivity vs frequency curve for proposed antenna.

Gain Vs. Frequency

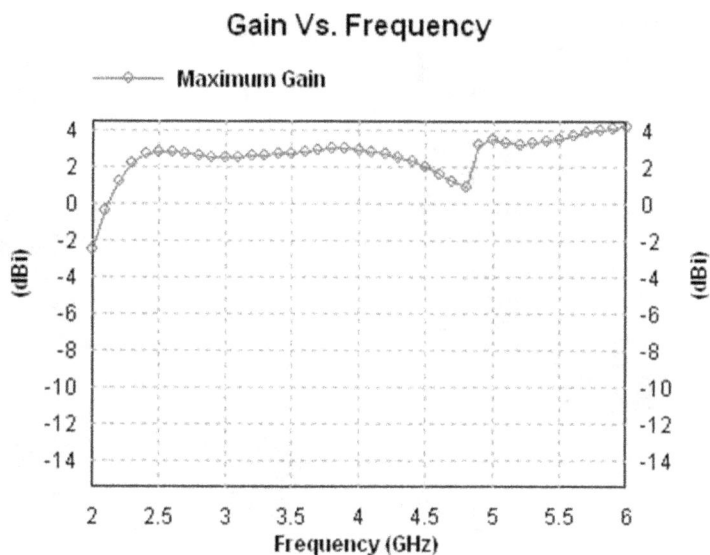

Figure 12.8 Gain curve for proposed antenna.

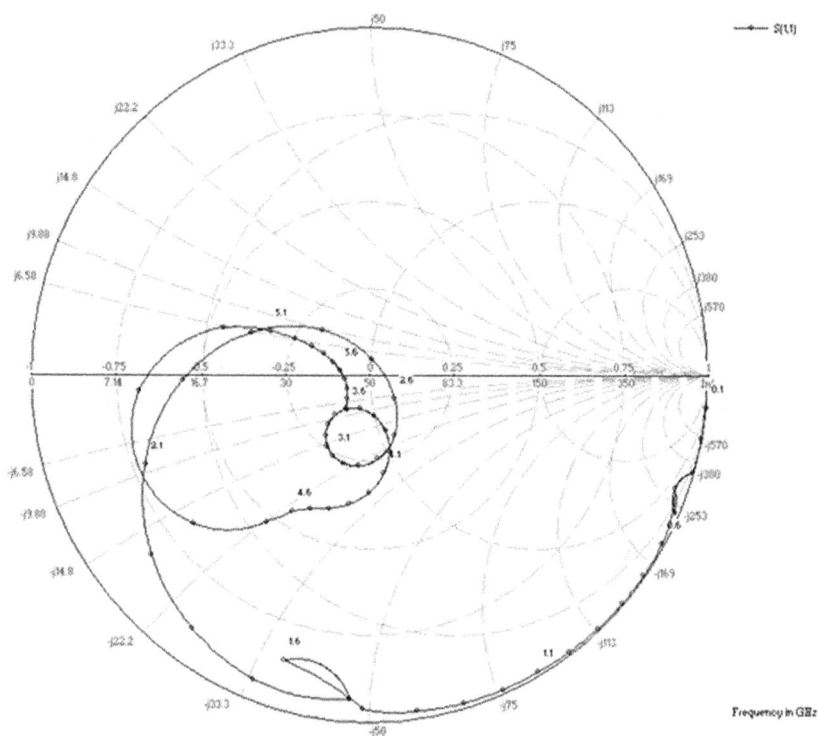

Figure 12.9 Smith chart for the proposed antenna.

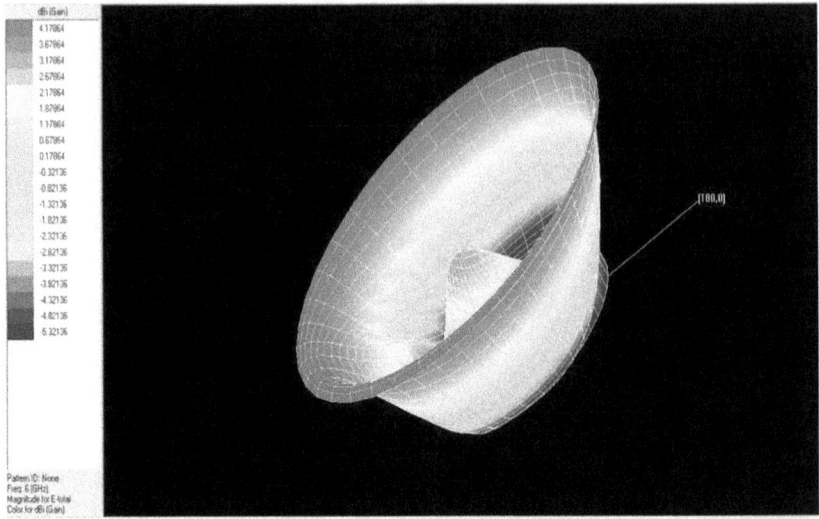

Figure 12.10 3D radiation pattern.

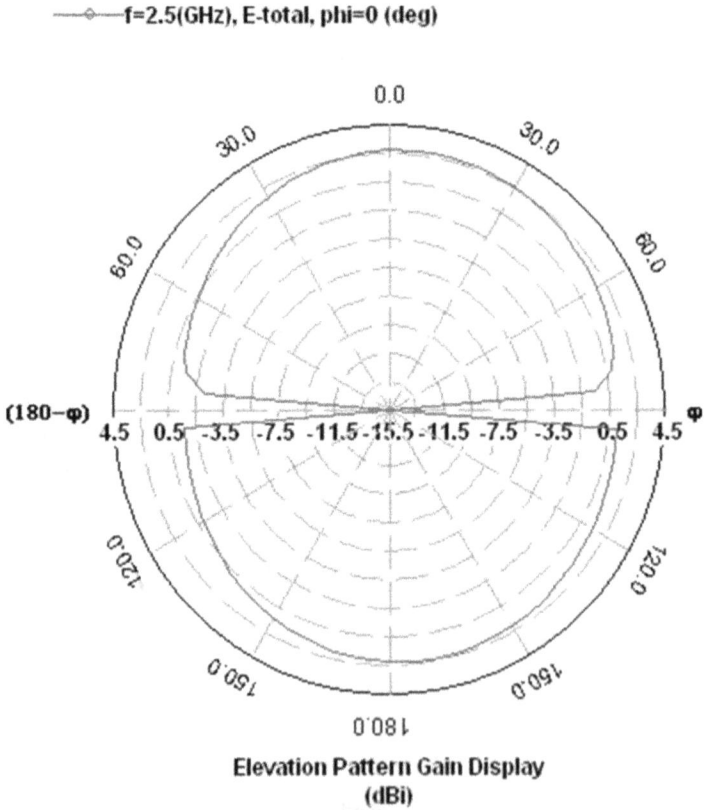

f=2.5(GHz), E-total, phi=0 (deg)

Elevation Pattern Gain Display
(dBi)

Figure 12.11 Radiation pattern of the proposed antenna.

12.5 VALIDATION

To ensure our graphically simulated results are reliable, we have also validated our results with experimental results obtained from other sources. The VSWR bandwidth is 73.36 %in a simulation conducted with the finite element method based software tool. The VSWR bandwidth spans in the simulation result are more compared to previous results. The antenna is validated using Agilent Network Anylazer N9923A, whose hardware result is approximately the same as the simulation results as shown in Figures 12.12–12.14.

The basic electrical characteristics of the proposed antenna are summarized in Table 12.2.

In Table 12.1 we proposed the dimension of the designed microstrip patch antenna. This is smaller than in the conventional antenna. In Table 12.2 we give the different parameter of the proposed antenna in which bandwidth is enhanced when compared with multiband planar antenna [6].

Figure 12.12 Top view of proposed antenna.

Figure 12.13 Ground plane of proposed antenna.

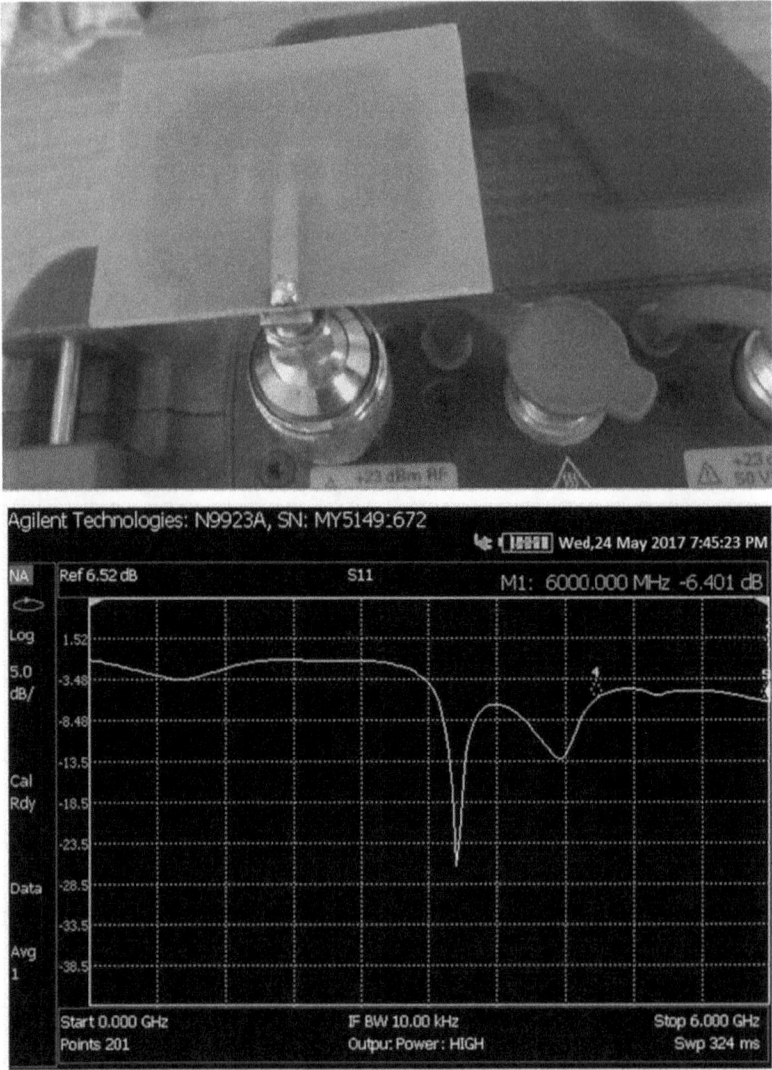

Figure 12.14 Hardware testing on Agilent N9923A.

Table 12.2 Basic electrical characteristics of the proposed antenna

S.no	Parameters	Value
1	Minimum S_{11}	−27 dB at 2.5 GHz
2	Bandwidth	73.36 %
3	VSWR	<2.0 between (2.4–4.1 and 5.1–6.3) GHz
4	Maximum directivity	6.2 dBi
5	Maximum gain	4.2 dBi

12.6 CONCLUSION

We have presented the design of the metamaterial antenna for bandwidth enhancement covering the 2 GHz–6 GHz frequency spectrum. The proposed metamaterial antenna has a bandwidth of approximately 73.36% with a steady radiation pattern within the frequency range. The proposed antenna for a center frequency have a good impedance matching (approximately 50 ohm). The proposed antenna is fabricated on FR4-substrate due to its small size and thickness. The simple microstrip feeding technique is used for the design.

REFERENCES

1. G.V. Eleftheriades, A.K. Iyer, and P.C. Kremer, "Planar Negative Refractive Index Media Using Periodically L-C Loaded Transmission Lines," *IEEE Trans. Microw. Theory Tech.*, vol. 50, pp. 2702–2712, 2002.
2. C. Caloz, and T. Itoh, "Positive/Negative Refractive Index Anisotropic 2-D Metamaterials," *IEEE Microw. Wirel. Compon. Lett.*, vol. 13, pp. 547–549, 2003.
3. C. J. Lee, K. M. K. H. Leong, and T. Itoh, "Composite Right/Left-Handed Transmission Line Based Compact Resonant Antennas for RF Module Integration," *IEEE Trans. Antennas Propag.*, vol. 54, no. 8, pp. 2283–2290, Aug. 2006.
4. T. G. Kim, and B. Lee, "Metamaterial-Based Compact Zeroth-Order Resonant Antenna," *Electron. Lett.*, vol. 45, no. 1, pp. 12–13, Jan. 2009.
5. J. Q. Huang, and Q. X. Chu, "Compact Epsilon Negative Zeroth Order Resonator Antenna with Higher Radiation Efficiency," *Microw. Opt. Technol. Lett.*, vol. 53, no. 4, pp. 897–900, Apr. 2011.
6. Dimitrios K. Ntaikos, Nektarios K. Bourgis, and Traianos V. Yioultsis, "Metamaterial-Based Electrically Small Multiband Planar Monopole Antennas," *IEEE Antennas And Wireless Propagation Letters*, vol. 10, pp. 963–966, 2011.
7. Cristhianne de Fátima Linhares de Vasconcelos, Maria Rosa Medeiros Lins de Albuquerque, Sandro Gonçalves da Silva, José de Ribamar Silva Oliveira, and Adaildo Gomes d'Assunção, "Full-Wave Analysis of Annular Ring Microstrip Antenna on Metamaterial," *IEEE Trans. Magn.*, vol. 47, no. 5, pp. 1110–1113, May 2011.
8. Puran Gour, "Bandwidth Enhancement of a Backfire Microstrip Patch Antenna for Pervasive Communication," *Int. J. Antennas Propag.*, vol. 7, pp. 1–9, 2014.
9. D. M. Pozar, *Microwave Engineering*, 2nd ed., John Wiley & Sons, 1998.
10. Constantine A. Balanis, *Antenna Theory Analysis and Design*, 3rd ed., John Wiley & Sons, 2005.

Chapter 13

Recent trends in 3D printing antennas

Geetam Richhariya, Rajesh Kumar Shukla,
Manish Sawale, and Nita Vishwakarma
Oriental Insititute of Science & Technology College, Bhopal, India

Nagendra Singh
Trinity College of Engineering & Technology, Karimnagar, India

13.1 INTRODUCTION

Over the past few decades, wireless communication has undergone rapid and transformative advancements, reshaping our perception of modern societies and altering the way we interact. Antennas, which are essential for sending and receiving radio frequency (RF) signals, are at the heart of wireless communication systems [1]. Researchers and businesses have been slavishly working on the 5G communication system, the next generation of communication systems, as the need for greater data speeds has increased over the past ten years. The introduction of 5G will promote the growth of the Internet of Things (IoT), leading to faster transmission rates, greater coverage, and more connections. With a growing array of devices incorporating embedded sensors, the volume of objects in communication networks is increasingexponentially. Consequently, the primary challenge faced by wireless communication lies in developing high-performance systems, particularly antennas, to meet these demands. Moreover, traditional communication frequency bands are facing constraints and overloads, posing challenges in providing the anticipated future data rates. As a result, higher frequency communication methods such as millimeter wave (MMW) and sub-millimeter wave (SMMW) must be used [2]. Limitations exist in high-frequency communication systems, especially in terms of the transmitters and receivers, where antennas are crucial. Highly directional antennas are necessary for effective communication in the millimeter wave (MMW) bands in order to combat free space attenuation and support large input-impedance bandwidths, strong radiation efficiency, and dependable radiation patterns. These requirements result in complicated structures, high levels of integration, lightweight materials, and the miniaturization of antennas, which put a burden on current production methods. Due to its beneficial characteristics, such as rapid prototyping, cheap cost, high material utilization, and the lack of any need for any special tools, 3D printing technology,

    DOI: 10.1201/9781003422440-13

also known as additive manufacturing, has attracted a lot of interest in both industry and research groups since the 1980s [3]. The layer-by-layer stacking technology used in 3D printing allows for a huge variety of structural possibilities, facilitating the construction of complicated shapes and small-scale manufacturing [4]. These benefits give 3D printing the chance to satisfy the needs of RF devices, such as antennas operating in MMW and higher frequency bands.

Making a digital model of the object that will be produced is the first step in the 3D printing process. A three-dimensional scanner such as computer-aided design (CAD) software, or photogrammetry technologies can all be used to create this virtual model. To create the model, photogrammetry integrates photos of the object obtained at various angles. An STL file, which holds the coordinates of triangulated sections representing the model's surfaces, is created once the 3D model is finished. The item is then divided into a succession of 2D cross-section layers along the z-axis using all of the ability of the 3D printer software to translate the STL files [5, 6]. This discrete strategy guarantees accurate layering. The 3D printer builds the desired thing layer by layer in the final stage. Figure 13.1 depicts a specific 3D printing procedure for reference.

Numerous research studies have investigated different kinds of 3D printed antennas, such as lens, dielectric resonator, and patch antennas, which are frequently made of metals, polymers, and ceramics. Many of these antennas exhibit dependable radiation patterns, high gain, exceptional radiation efficiency, and ultrabroad band capabilities, making them appropriate for use in satellites, mobile communication, remote sensing, and other MMW band applications. Despite major advancements in the 3D printing of antennas, additional research is still needed to develop and create more complex structures for the MMW and terahertz (THz) bands. Achieving this goal involves investigating different techniques and developing diverse materials to meet the specific challenges of these frequency bands. The chapter emphasizes the key advantages of 3D printing while addressing the challenges involved in

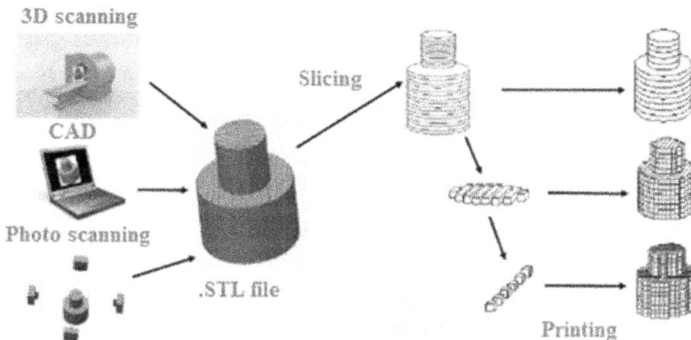

Figure 13.1 The process of 3D printing.

fabricating antennas using this technology [7]. The study is organized into five sections. The first section introduces the background and significance of the research.

13.1.1 Printing technology (3D) for antennas

Materials extrusion, binder jetting, material jetting, directed energy deposition, sheet lamination, powder bed fusion, and vat photopolymerization are only a few of the many 3D printing processes that have been developed [8]. Fused deposition modelling (FDM), which is related to material extrusion, stereolithography apparatus (SLA), which is related to vat photopolymerization, polyjet, which is related to material jetting, and selective laser melting (SLM), which is related to powder bed fusion, are among the commonly used technologies in the field of antenna manufacturing. Additionally, binder jetting (BJ) has been mentioned as a method for producing antennas [9].

13.1.2 Modelling of fused deposition

Utilizing material extrusion, the technology melts filament using an extrusion nozzle and layers it to produce the desired objects. Polylactic acid (PLA), polycarbonate (PC), and acrylonitrile butadiene styrene (ABS) are among the polymers that are frequently utilized in this process. However, material extrusion (FDM) is typically only used to fabricate plastic parts due to material and process restrictions [10]. This poses a challenge when manufacturing metallic antennas, as additional post-processing steps, like metallization for conductors, are needed. Nevertheless, a recent development, called "Electrification," offers a promising solution. A more durable, commercially accessible filament called "Electrification" is made of a special metal–polymer combination mostly made of nonhazardous copper and biodegradable polyester. With Electrification, the complete antenna construction may be 3D-printed without the need for additional methods for conducting elements. Due to its low cost, FDM is used extensively, but its accuracy, which is based on the size of the extruder nozzle and how precisely it moves, is subpar. With typical nozzle widths of 200 mm or greater and surface roughness ranges of 10 to 50 mm, FDM is better suited to the production of low-frequency antennas [11].

13.1.3 Stereolithography

The main method used by Selective Laser Additive is photopolymerization. A light source is used in this technique to selectively harden UV-sensitive polymers, providing the building blocks for 3D printing tiny components. SLA is one of the most accurate 3D printing techniques available today, thanks to its exact laser light spot size and x-y-z mobility. SLA printers can attain resolutions of 20m or even less, whilst the majority of commercial 3D printers have

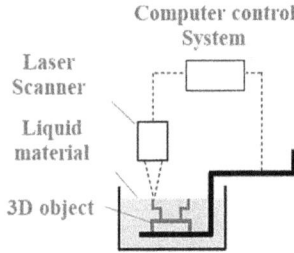

Figure 13.2 Hull's stereolithography method.

a minimum feature size [8]. The SLA is appropriate for operations in the MMW bands because it can produce objects with flat surfaces. It is important to note that SLA can be divided into two categories based on the materials used: polymer SLA and ceramic SLA (CSLA). Applications for high-filtering and RF packaging have seen successful using polymer SLA. Due to its lossy nature at microwave frequencies, however, it is not appropriate for low-loss MMW applications. Hull's stereolithography system is shown in Figure 13.2.

By contrast, CSLA involves post-printing debinding and sintering procedures that eliminate the polymer percentage and leave only the ceramic components. Because of this, CSLA is necessary for producing dielectric antennas, particularly lens antennas and dielectric resonator antennas (DRAs), as it enables the manufacture of all-ceramic components. It has helped CSLA become more widely used in the field of microwaves and 5G technology that researchers have extensively examined microwave ceramics with high-quality factors and low dielectric loss in the past [12, 13].

13.1.4 Polyjet technology for 3D printers

With the use of a jetting head, liquid materials are sprayed on a surface using polyjet technology. These materials immediately solidify under UV light to form thin layers. The method uses photopolymerization to produce many kinds of polymers simultaneously utilizing various printer heads. This makes it possible to print overhanging structures with gel-type polymers as support materials that are then easily removed; for instance, by utilizing support materials that are water-soluble and washable [14]. Polyjets provide a number of benefits, including the quick and inexpensive production of polymer components with smooth surfaces and complicated shapes. However, a metallization procedure is needed to include conductors. Polyjet has previously produced successful electromagnetic bandgap structures and several kinds of dielectric antennas [15]. Polyjet is a viable technology for 3D printed antennas working at MMW, SMMW, and THz frequencies because of its capabilities. Additionally, Polyjet enables the cost-effective production of antennas with complex structures while maintaining accuracy, convenience, and quicker fabrication processes when compared to older techniques.

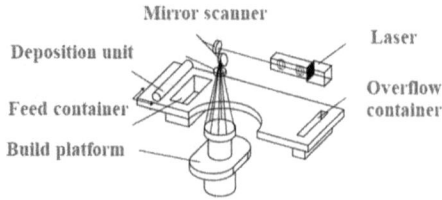

Figure 13.3 Schematic layout of DMLS setup.

13.1.5 Direct laser sintering of metal

Indirect metal laser sintering (IMLS) uses a powdered material that can be either metal, ceramic, or polymer. The process involves the laser beam selectively melting the powder particles on the printing plane, effectively transforming them into solid parts. However, the resolution obtained, which is affected by the laser spot, the size of the powder grain, and the movement of the laser, is considered unsatisfactory as it typically reaches only tens of microns. Certain non-melted powder particles were discovered during a subsequent scanning electron microscopy (SEM) analysis, adding to the roughness of the manufactured component. Nevertheless, the versatility of DMLSlies in its ability to print various materials, including metallic powders, making it possible to directly produce metal antennas without the need for post-processing [16, 17]. Additionally, antennas fabricated using DML Soffer advantages such as high densification, the absence of holes, and exceptional strength. As a result, DMLS has become a widely-used method for fabricating microwave antennas. Figure 13.3 shows a schematic layout of DMLS setup describing various parts.

13.1.6 Binder jet 3D printing

In order to bind the powder particles together and produce a 3D shape, the binder jetting (BJ) procedure includes the selective injection of liquid binder onto the top of a powder bed [18]. It operates at a significantly lower temperature than sintering or melting procedures, negating the requirement for shielding gas, preheating, or a vacuum chamber. To improve their mechanical qualities, however, the as-printed parts frequently need post-processing [19]. Similar to melting techniques, nozzle movement and powder size also limit the accuracy of BJ, with the smallest feature size currently practicable being 0.1 mm. Metals, ceramics, polymers, and biomaterials are just a few of the several materials that can be used with BJ [20]. BJ hasnot received as much attention as other 3D printing methods in the field of antenna fabrication, and metals are typically the material of choice. BJ does, however, have potential for high-speed, low-cost fabrication [21]. Various elements utilized in 3D-printed binder jets are described in Figure 13.4.

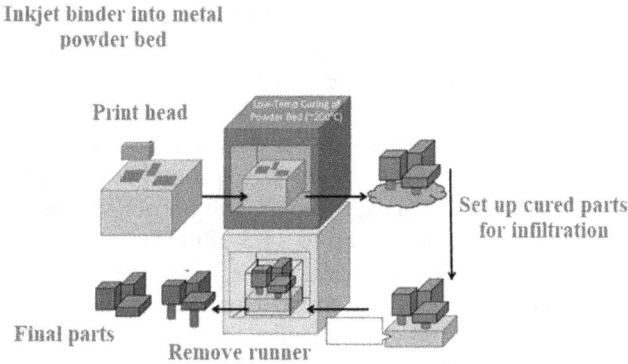

Figure 13.4 Various parts binder jet 3D printed.

For the manufacture of antennas, Polyjet and binder jetting (BJ) have also been used in some situations [22]. The wide variety of printing techniques is essential for developing various antenna systems, advancing communication systems, and advancing other disciplines.

13.2 PRINTED 3D ANTENNAS

Antennas have been made from a variety of materials, including thermoplastics, resins, ceramics, and metal powder. Each of these materials comes with its own set of advantages and disadvantages, some of which are minor, while others are more significant. The following categories best describe 3D printed antennas in terms of the materials they are made of: 1) Metallic antennas, 2) Polymer antennas, 3) Ceramic antennas printed directly with conductive materials or printed with dielectric materials and then metalized, 4) Composite material antennas primarily involving polymer matrices with ceramics, and 5) Multi-material integrated antennas, where different parts are made of conductors and dielectric materials [23]. The 3D printing methods available varies depending on the materials used. Some methods prove less effective and less commonly used due to challenges related to design complexity and material handling. The use of various 3D printing processes for various materials will be thoroughly covered in the sections that follow.

13.3 LENS ANTENNAS

Polymers are frequently used in the manufacture of antennas because they are easier to print than conductive materials. Therefore, it turns out that 3D printing is the best technique for creating dielectric antennas, such as lenses and dielectric resonator antennas (DRAs). Lens antennas represent a viable alternative for high-frequency and highly directional applications. A variety of 3D printing techniques have been used to successfully manufacture lens antennas throughout a wide frequency range, from microwaves to MMW bands and even up to THz [24]. The Luneburg lens (LL), which can offer a wide bandwidth without aberrations while simultaneously reducing system complexity and beam scanning costs, has drawn interest in radar and satellite systems. Prototypes of these designs were printed using techniques such as polyjet and polymer materials, demonstrating promising performance in the Ka-band, with gains reaching up to 21.3 dBi [25].

Additionally, a simplified spherical LL was successfully printed using polymer materials and covered the entire Ka-band with a gain of up to 21.2 dBi [26]. However, it's essential to consider the resolution of 3D printers as it can greatly impact antenna performance. Failure to achieve the desired results may occur due to a large laser spot [27]. To overcome resolution limitations, masked stereolithography (MSLA) was employed to successfully print the LL with higher resolution using "Grey Touch" UV-curable resins [28]. 3D printing has opened up exciting possibilities for the manufacturing of dielectric antennas, particularly lenses and DRAs, across various frequencies. The advancements in this field continue to drive innovation and improve antenna performance.

This highlighted the need for additional advancements in 3D printing technology and was mostly caused by improper lens printing and assembly as well as the greater dielectric constant and dielectric loss of the superglue utilized in the procedure. The shape and proportions of the lens inner structure, as well as the choice of material, must be carefully taken into account when designing MMW gradient index lenses. A number of geometrical changes have been made to the Luneburg lens employing transformation optics coordinate transformations in order to adapt it for practical uses. The flat LL, which is frequently built as a ring structure with numerous layers, is an optical transformation of the Luneburg lens. The air volume percentage was changed from the innermost layer to the outermost layer with a lower PLA in order to obtain the necessary effective permittivity values [29].

Accurately calculating the loss tangent and dielectric constant is essential for increasing the gain of the FZPL. In order to improve the lens design for microwave applications, two additional FZPL antennas were 3D printed using customized dielectric materials [30]. The Fresnel lens" overall performance and efficiency were improved by using a revolutionary unit cell architecture that made it easier to develop and install. Recent advancements in conformal antenna design have utilized transformation optics and multi-objective topological optimization to create printable antennas [31]. Additionally, 3D printing has been used to create antennas for specialized radios, including an 8-beam gradient-index lens made for mobile adhoc networks with 360° coverage and several functions. Its symmetric construction allows it to radiate with an ideal 45° beam width in each of its eight sectors and gives it the flexibility to switch to a multi-beam mode that spans eight directions [32]. Small antennas that are simple to insert into devices are crucial in the context of future 5G mobile communication. Lens antennas are significant in the THz band as well as the microwave and MMW bands, where they are essential parts of THz communication systems. THz lenses have been produced using sub-micron precision micro SLA 3D printing [33]. Additionally, PA6 has been used to print affordable diffractive THz lenses that operate at 0.625 THz for imaging and focusing in THz systems [34].

Dielectric resonator antennas (DRAs), which differ from lens antennas in that they are lightweight, compact, and highly radiation efficient, are also important. They have numerous uses in UWB antennas, MMW antennas, and MIMO antennas [35].

A 3D-printed plasma DRA made of resin has been developed, allowing for changeable gain and reconfigurable radiation patterns by modifying the excitation state of the plasma. This configuration offers a cost-efficient solution for multifunction WLAN antennas.

13.4 METAL 3D PRINTING

Metallic antennas created through 3D printing methods offer innovative solutions in modern antenna design. This technology enables the precise and customizable manufacturing of antennas with complex geometries, providing enhanced performance and functionality across various applications. By utilizing metallic materials and advanced 3D printing processes, these antennas can be tailored to specific frequency bands, radiation patterns, and impedance requirements, making them a versatile choice for diverse communication systems, including wireless communication, radar, and satellite applications. The sector of telecommunications may undergo a revolution as a result of the capacity to quickly prototype and improve metallic antennas through the use of 3D printing [36].

13.4.1 Direct metal 3D printing

The exploration and development of printed metallic antennas presents a promising direction in antenna design. Classic antenna configurations typically require conducting elements for resonance, but using poor conductors for 3D printed electrical circuits results in significant losses, making them unsuitable for antenna applications. In contrast, metallic 3D printed parts offer stronger mechanical robustness and lower electrostatic discharge compared to dielectric antennas. Through techniques like selective laser melting (SLM) or direct metal laser sintering (DMLS), it is possible to print pure metallic antennas directly from materials like steel, aluminium, or titanium. For MMW and SMMW applications, horn antennas and lens antennas are popular options. The horns' outside surfaces were cleaned through post-processing, while their inner surfaces were created to preserve the sound structure of the ridges and steps. Even though surface roughness can be lessened through post-processing, further study is necessary due to the complicated structure of antennas. Additionally, SLM printing technology was used to create X-band horn antennas. Such antennas demonstrated excellent results, achieving an 18 dBi directivity at 7.9 GHz and a thickness of 1 mm, with measurements aligning with the modeled predictions [37]. The advancements in 3D printed metallic antennas hold great potential for advancing antenna technology and performance in various communication applications.

Higher-gain horns were also successfully 3D printed using Cu-Sn via SLM, and manual polishing was used as the post-processing technique due to its benefits in roughness reduction and cheap compatibility cost [38]. AlSi10Mg was a standout material in the DMLS process, offering strength, lightweight qualities, and excellent electrical characteristics, making it appropriate for the production of RF components. Additionally, the DMLS process permits the use of both pure metals and metal alloys without sacrificing their material properties. Axial corrugated horns can be printed using MMW 3D metal technology since they have acceptable tolerance requirements and a matching bandwidth of about 20 dBi for the complete Ka-band. Fractal antennas, based on self-similar iterations in different directions and scales, are ideal candidates for 3D printing due to their complex features. A copper layer was deposited on the antenna to increase surface conductivity. In all cases, the fractal antennas' unique architecture reduced the amount of material needed by more than 75%, while the 3D printing technique added mechanical strength, notably at the junctions where several fractal element repeats meet.

The difficulty in creating metallic elements is one of the key drawbacks of 3D printing technology, making it challenging to print conductive materials for electromagnetic (EM) applications [39]. When coated with copper tape or conductive copper paint, the antenna made from conductive filaments outperformed horns printed with PLA in terms of gain and bandwidth and required no post-print processing. The overall cost of producing antennas remained lower despite having greater production costs than metal-based antennas with comparable properties. DMLS technology has opened up various applications for waveguide-based passive microwave components, antennas, and antenna arrays [40]. Waveguide-based parts and antennas that were previously difficult to manufacture can now be made thanks to DMLS. However, there are still certain problems with direct printing metallic antennas that need to be resolved. Notably, the surface roughness of 3D-printed metals is lower than that of 3D-printed dielectrics [41]. The functionality of an antenna component is greatly influenced by a number of important aspects, including printing resolution, structural constraints, and surface roughness. The MMW/THz antennas' short wavelength necessitates complex geometrical features, which places a heavy burden on the production processes. Enhancing printing resolution and reducing surface roughness are essential steps to prepare antennas for higher frequencies. Additionally, addressing the challenge of printing conductors necessitates the development of new conductive materials specifically tailored for 3D printing.

13.5 CERAMIC ANTENNAS FOR 3D PRINTING

Due to their distinctive characteristics and possible applications, 3D printed ceramic antennas have attracted a lot of attention. Ceramics offer advantages such as high temperature stability, low dielectric loss, and excellent mechanical properties, making them suitable for various antenna designs. Ceramic antennas can be fabricated using processes like SLA or polyjet, allowing for precise and complex geometries. The ability to 3D print ceramic antennas opens up possibilities for customizing antenna designs and optimizing their performance for specific frequencies and applications. As the technology continues to advance, 3D printed ceramic antennas are expected to play a crucial role in various industries, including communication systems, aerospace, and healthcare [42].

13.6 3D PRINTED CERAMIC ANTENNAS

Microwave dielectric ceramics, alongside polymer materials, are crucial in wireless communication, particularly in 5G applications. Various ceramic materials, such as Al_2O_3 and $MgTiO_3$, have been used to fabricate LL antennas [43]. The CSLA process is commonly employed for accurate manufacturing of dielectric ceramic antennas. In order to attain tunable dielectric constants, composite ceramics are frequently utilized, which makes them useful for RF devices such as lens antennas and filters. 3D printing with ceramic materials allows for the creation of anisotropic dielectrics, providing more design possibilities for DRAs. This versatility in adjusting microwave EM characteristics by mixing different materials opens upnew prospects for the engineering of antennas. Integrated lens antennas (ILAs) fabricated with Al_2O_3 via CSLA have shown improved radiation performance, indicating the potential benefits of higher resolution and smaller-sized elements in 3D printed objects. The use of microwave dielectric ceramics is expected to expand with the rise of 5G technology, particularly in aerospace applications where they offer small size, low weight, and high quality. For instance, ceramic antenna manufacturing utilizes CSLA with high resolution, enabling the production of intricate structures. These ceramic materials possess low dielectric loss, a high microwave quality factor, and a near-zero temperature coefficient of resonant frequency. These beneficial characteristics make it easier to integrate and miniaturize RF components for 5G applications [44]. Ceramics have a wider variety of dielectric permittivities and more

dependable mechanical qualities than polymers, enabling better control of
EM material parameters. The outstanding radiation performance attained
by ceramic antennas has been demonstrated in several researches. The use
of printable ceramics is expected to increase the range of applications for
3D printed antennas.

13.7 COMPOSITE MATERIAL IN 3D PRINTING

Using a polymer matrix composite technique, the constrained dielectric constant range is addressed, involving the mixing of a 3D printable polymer
with nanoparticles and different ratios of ceramics [45]. In order to support IoT and 5G applications, Chietera et al. developed a novel design for
a wideband DRA using a PLA-based 3D-printable filament material loaded
with 17.5% $BaTiO_3$ as the primary component of the resonating structure
[46]. The issue of limited dielectric constant range is resolved and the small
size demands of 5G communication systems are met by the use of polymer
matrix composites. Additionally, Liu et al. showed that printing antenna substrates using $BaTiO_3$ PLA-based filaments can reduce antenna size by about
40% in terms of the surface ratio of the radiating element [47]. Additionally,
by employing a simple meandering technique, the antenna size can be further reduced. Metamaterials, which are unnaturally produced intermittent
structures with soleassets not found in natural materials, offer various applications, including cloaking and wireless communication. Polymer matrix
composites were created using ABS or PLA as the base material, with the
addition of ceramic particles. The resulting composite filament exhibited
high permittivity, while the pure polymer material had low permittivity. A
dielectric meta-material with anisotropic characteristics was successfully
printed by carefully arranging high- and low-permittivity filament materials in the printed coupons. It is now possible to develop novel materials
with a larger variety of EM microwave properties by using variable loading
fractions of composite materials, enabling additive production of complex
structures with spatially varied EM features [48]. This creates possibilities
for the creation of innovative antenna structures in cutting-edge fields like
5G communications.

13.8 CONCLUSION

A variety of materials have been utilized in designing and fabricating antennas to achieve specific EM properties and desired radiation performance.
Extensive research has made polymer antennas suitable for applications in
microwave and THz bands, providing lightweight solutions for 5G communication requirements. However, they lack the mechanical robustness
of metallic antennas, which, though fabricated through direct 3D printing,

offer stronger mechanical durability but are limited in high-frequency applications due to their weight and surface roughness. Metallic antennas produced through metallization after 3D printing offer cost and weight advantages over full-metal antennas but are less mechanically robust and unsuitable for harsh environments or high-power scenarios. In contrast to polymer and metal alternatives, ceramic materials provide a wider range of dielectric permittivities and more stable mechanical qualities, allowing for smaller antennas. Composite materials show promise in addressing limited dielectric permittivity, and their use can reduce antenna size by incorporating higher dielectric permittivity composites. Multimaterial integrated antennas offer diverse material choices, reducing costs and weight, but may pose challenges in mechanical robustness when combining different parts. The antenna's type and structure significantly impact its radiation performance.

In terms of lens antennas (LL), they are favored for radar and satellite applications, offering high gain, low subflap, and broadband characteristics. Among several 3D printing technologies, LL lenses made with CSLA have the highest gain and highest resolution. LL lenses with polymer materials manufactured via FDM provide significant gain at high frequencies. In contrast to polymer or ceramic LLs, integrated lens antennas (ILAs) often consist of more than one material, comprising an antenna and an LL. However, gains more than 10 dBi are frequently attained by LL antennas produced utilizing a variety of 3D printing procedures and materials, demonstrating their high-gain nature. In order to create high-performance systems, it is necessary to thoroughly compare the benefits and drawbacks of various materials and 3D printing technologies. The overall performance of antennas is determined by material, structure, and printing accuracy.

13.9 CHALLENGES AND PROSPECTS

The potential of 3D printing in revolutionizing manufacturing lies in its ability to quickly produce complex structures with greater design flexibility compared to traditional methods. However, there are several obstacles to overcome before achieving reliable production of advanced-functioning antennas using 3D printing. The current interest in the prospects of 3D printing technology in 5G applications has prompted this chapter's review of the state-of-the-art 3D printed antennas across different frequency bands, categorized by material types: polymer, metallic, ceramic, composite, and multimaterial integrated antennas. The chapter thoroughly assesses the advantages and disadvantages of each antenna type, considering factors such as cost and weight. In-depth comparisons of the resolution, printing speed, and printing size of several 3D printing technologies relevant to antennas are also provided. The review also explores the challenges in this field and proposes future directions for the development of 3D printed antennas to achieve fully functional microwave systems. Overall, the findings suggest that 3D printing holds promise as a viable option for mobile communications, offering valuable guidance for advanced research in this area.

REFERENCES

1. D. Helena, A. Ramos, T. Varum, et al, "Inexpensive 3D-Printed Radiating Horns for Customary Things in IoT Scenarios," *Proceedings of the 14th European conference on antennas and propagation (EuCAP)*, IEEE, 2020.

2. S. Alkaraki, Y. Gao, S. Stremsdoerfer, et al, "3D Printed Corrugated Plate Antennas with High Aperture Efficiency and High Gain at X-band and Ka-Band," *IEEE Access*, vol. 8, pp. 30643–30654, 2020.

3. R. Colella, F. P. Chietera, F. Montagna, et al, "On the Use of Additive Manufacturing 3D-Printing Technology in RFID Antenna Design," *Proceedings of the IEEE International Conference on RFID Technology and Applications (RFID-TA)*, IEEE, 2019.

4. S. Kingsley, *Advances in Handset Antenna Design*, RFDes, pp. 16–22, 2005.

5. Wu Z, Kinast J, Gehm M, et al., "Rapid and Inexpensive Fabrication of Terahertz Electromagnetic Bandgap Structures," *Opt. Express*, vol. 16, no. 21, pp. 16442–16451, 2008.

6. Yuanyuan Xu, Xiaoyue Wu, Xiao Guo, Bin Kong, Min Zhang, Xiang Qian, Shengli Mi, and Wei Sun, "The Boom in 3D-Printed Sensor Technology," *Sensors*, vol. 17, pp. 1166, 2017.

7. C. A. Balanis, "Antenna Theory: Are View," *Proc. IEEE*, vol. 80, no. 1, pp. 7–23, 1992.

8. W. Hong, Z. Jiang, C. Yu, et al. "The Role of Millimeter-Wave Technologies in 5G/6G Wireless Communications," *IEEE J. Microw.*, vol. 1, no. 1, pp. 101–122, 2021.

9. H. Guo, D. Zhou, C. Du, et al., "Temperature Stable Li2Ti0.75 (Mg1/3Nb2/3) 0.25O3-Based Microwave Dielectric Ceramics with Low Sintering Temperature and Ultra-Low Dielectric Loss for Dielectric Resonator Antenna Applications," *J. Mater. Chem. C*, vol. 8, no. 14, pp. 4690–4700, 2020.

10. D. K. Ahn, S. M. Kwon, and S.H. Lee, "Expression for Surface Roughness Distribution of FDM Processed Parts," *Proceedings of the International Conference on Smart Manufacturing Application*, IEEE, 2008.

11. F.P. Melchels, J. Feijen, and D. W. Grijpma, "A Review on Stereo Lithography and its Applications in Biomedical Engineering," *Biomaterials*, vol. 31, no. 24, pp. 6121–6130, 2010.

12. J. Huang, Q. Qin, and J. Wang, "A Review of Stereo Lithography: Processes and Systems," *Processes*, vol. 8, p. 1138, 2020.

13. S. Moreno-Rodríguez, M. A. Balmaseda-Márquez, J. Carmona-Murillo, et al. "Polarization-Insensitive Unit Cells for a Cost-Effective Design of a 3-D-Printed Fresnel-Lens Antenna," *Electronics*, vol. 11, no. 3, p. 338, 2022.

14. A. Goulas, G. Chi-Tangyie, D. Wang, et al., "Microstructure and Microwave Dielectric Properties of 3D Printed Low Loss $Bi_2 Mo_2 O_9$ Ceramics for LTCC Applications," *Appl. Mater. Today*, vol. 21, p. 100862, 2020.

15. D. Wang, J. Chen, G. Wang, et al., "Old Sintered LiMgPO4 Based Composites for Low Temperature Co-Fired Ceramic (LTCC) Applications," *J. Am. Ceram. Soc.*, vol. 103, no. 11, pp. 6237–6244, 2020.

16. H. Xiang, C. Li, Y. Tang, et al., "Two Novel Ultralow Temperature Firing Microwave Dielectric Ceramics LiMVO$_6$ (M = Mo, W) and Their Chemical Compatibility with Metal Electrodes," *J. Eur. Ceram. Soc.*, vol. 37, no. 13, pp. 3959–3963, 2017.

17. B. Zhang, Y. Li, and Q. Bai, "Defect Formation Mechanisms in Selective Laser Melting: Review," *Chin. J. Mech. Eng.*, vol. 30, pp. 515–527, 2017.

18. Y. Liu, L. Peng, and W. Shao, "An Efficient Knowledge-Based Artificial Neural Network forthe Design of Circularly Polarized 3D-Printed Lens Antenna," *IEEE Trans. Antennas Propag.*, vol. 70, no. 7, pp. 5007–5014, 2022.

19. M. Liang, W.R. Ng, K. Chang, et al., "A 3-D Luneburg Lens Antenna Fabricated by Polymer Jetting Rapid Prototyping," *IEEE Trans. Antennas Propag.*, vol. 62, no. 4, pp. 1799–1807, 2014.

20. Z. Wu, M. Liang, W. R. Ng, et al., "Terahertz Horn Antenna Based on Hollow–Core Electromagnetic Crystal (EMXT) Structure," *IEEE Trans. Antennas Propag.*, vol. 60, no. 12, pp. 5557–5563, 2012.

21. F. Zhou, W. Cao, B. Dong, et al., "Additive Manufacturing of a Three-Dimensional Terahertz Gradient-Refractive Index Lens," *Adv. Opt. Mater.*, vol. 4, no. 7, pp. 1034–1040, 2016.

22. Eyob Messele Sefene, "State-of-the-Art of Selective Laser Melting Process: A Comprehensive Review," *J. Manuf. Syst.*, vol. 63, pp. 250–274, Apr. 2022.

23. S. M. Gaytan, M. A. Cadena, H. Karim, et al., "Fabrication of Barium Titanate by Binder Jetting Additive Manufacturing Technology," *Ceram. Int.*, vol. 41, no. 5, pp. 6610–6619, 2015.

24. M. Ziaee, N. B. Crane, "Binder Jetting: A Review of Process, Materials, and Methods," *Addit. Manuf.*, vol. 28, pp. 781–801, 2019.

25. Z. Li, Z. Chen, X. Liu, et al. "A 3D Printed Plasma Dielectric Resonator Antenna," *Proceedings of the IEEEMTT-S International Conference on Numerical Electromagnetic and Multiphysics Modeling and Optimization (NEMO)*, IEEE, 2020.

26. C. Yang, Y. Xiao, and K. W. Leung, "A 3D-Printed Wideband Multilayered Cylindrical Di-Electric Resonator Antenna with Air Layers," In *Proceedings of the IEEE Asia-Pacific Microwave Conference (APMC)*, IEEE, 2020.

27. R. K. Luneburg, "Mathematical Theory of Optics," *Am.J. Phys.*, vol. 34, no. 1, pp. 80, 1966.

28. Y. Guo, Y. Li, J. Wang, et al., "A 3D Printed Nearly Isotropic Luneburg Lens Antenna for Millimeter-Wave Vehicular Networks," *IEEE Trans Veh Technol*, vol. 71, no. 2, pp. 1145–1155, 2022.

29. Y. Li, L. Ge, M. Chen, et al., "Multi-Beam 3D Printed Luneburg Lens Fed by Magneto-Electric Dipole Antennas for Millimeter-Wave MIMO Applications," *IEEE Trans Antennas Propag*, vol. 67, no. 5, pp. 2923–2933, 2019.

30. Z. Song, H. Zheng, Y. Li, et al., "Investigation of Axial Mode Dielectric Helical Antenna," *IEEE Trans. Antennas Propag.*, vol. 70, no. 5, pp. 3806–3811, 2021.

31. J. Huang, S. J. Chen, Z. Xue, et al., "Wideband Endfire 3-D-Printed Dielectric Antenna with Designable Permittivity," *IEEE Antennas Wirel. Propag. Lett.*, vol. 17, no. 11, pp. 2085–2089, 2018; P. Kadera, A. Jimenez-Saez, T. Burmeister, et al., "Gradient-Index-Based Frequency-Coded Retroreflective Lenses for mm-Wave Indoor Localization," *IEEE Access*, vol. 8, pp. 212765–212775, 2020.

32. A. S. Gutman, "Modified Luneberg Lens," *J. Appl. Phys.*, vol. 25, no. 7, pp. 855–859, 1954.

33. A. Demetriadou, and Y. Hao, "Slim Luneburg Lens for Antenna Applications," *Opt. Express*, vol. 19, no. 21, pp. 19925–19934, 2011.

34. Gilmore J., Kotzék, "SLM3D-Printed Horn Antenna for Satellite Communications at X-band," *Proceedings of the IEEE-APS Topical Conference on Antennas and Propagation in Wireless Communications (APWC)*, IEEE, 2019.

35. B. Zhang, Z. Zhan, Y. Cao, etal., "Metallic 3-D Printed Antennas for Millimeter- and Submillimeter Wave Applications," *IEEETrans. Terahertz Sci. Technol.*, vol. 6, no. 4, pp. 592–600, 2016.

36. R. Colella, F. P. Chietera, F. Montagna, et al., "On the Use of Additive Manufacturing 3D-Printing Technology in RFID Antenna Design," *Proceedings of the IEEE International Conference on RFID Technology and Applications (RFID-TA)*, IEEE, 2019.

37. T.H. Chio, G. Huang, and S. Zhou, "Application of Direct Metal Laser Sintering to Waveguide-Based Passive Microwave Components, Antennas, and Antenna Arrays," *Proc. IEEE*, vol. 105, no. 4, pp. 632–644, 2017; Y. Wu, D. Isakov, and P. S. Grant, "Fabrication of Composite Filaments with High Dielectric Permittivity for Fused Deposition 3DPrinting," *Materials*, vol. 10, no. 10, p. 1218, 2017 (Basel).

38. B. Khatri, K. Lappe, M. Habedank, et al., "Fused Deposition Modeling of ABS-Barium Titanate Composites: A Simple Route Towards Tailored Dielectric Devices," *Polymers*, vol. 10, no. 6, p. 666, 2018 (Basel).

39. A. Dorle, R. Gillard, E. Menargues, et al., "Additive Manufacturing of Modulated Triple-Ridge Leaky-Wave Antenna," *IEEEAntennasWirel.Propag. Lett.*, vol. 17, no. 11, pp. 2123–2127, 2018.

40. H. J. M. Odiaga, May Medina, and S. A. Navarro, "An Implemented 3D Printed Circular Waveguide Antenna for K Band Applications," *Proceedings of the IEEE MTT-S Latin America Microwave Conference (LAMC 2018)*, IEEE, 2018.

41. Y. Shi, and L. M. Bhowmik, "An Effective Method for Fabrication of 3-D Dielectric Materials Using Polymer-Ceramic Composites," *Proceedings of the IEEE Antennas and Propagation Society International Symposium (APSURSI)*, IEEE, 2014.

42. D. C. Lugo, R. A. Ramirez, J. Castro, et al., "3D Printed Multilayer mm-Wave Dielectric Rod Antenna With Enhanced Gain," *Proceedings of the IEEE International Symposium on Antennas and Propagation & USNC/URSI National Radio Science Meeting*, IEEE, 2017.

43. S. Alkaraki, Y. Gao, S. Stremsdoerfer, et al., "3D Printed Corrugated Plate Antennas With High Aperture Efficiency and High Gain at X-Band and Ka-Band," *IEEE Access*, vol. 8, pp. 30643–30654, 2020.

44. S. Alkaraki, A. S. Andy, Y. Gao, et al., "Compact and Low-Cost 3-D Printed Antennas Metalized Using Spray-Coating Technology for 5G mm-Wave Communication Systems," *IEEE Antennas Wirel. Propag. Lett.*, vol. 17, no. 11, pp. 2051–2055, 2018.

45. J. J. Adams, E. B. Duoss, T. F. Malkowski, et al., "Conformal Printing of Electrically Small Antennas on Three-Dimensional Surfaces," *Adv.Mater.*, vol. 23, no. 11, pp. 1335–1340, 2011.

46. M. Liang, C. Shemelya, E. Mac Donald, et al., "Fabrication of Microwave Patch Antenna Using Additive Manufacturing Technique," *Proceedings of the USNC-URSI Radio Science Meeting (Joint with AP-S Symposium)*, IEEE, 2014.

47. Y. C. Toy, P. Mahouti, F. Güneş, et al., "Design and Manufacturing of an X-Band Horn Antenna Using 3-D Printing Technology," *Proceedings of the 8th International Conference on Recent Advancesin Space Technologies (RAST)*, IEEE, 2017.

48. D. Shamvedi, O. J. McCarthy, E. O'Donoghue, et al., "Improving the Strength-to-Weight Ratio of 3-D Printed Antennas: Metal versus Polymer," *IEEE Antennas Wirel. Propag. Lett.*, vol. 17, no. 11, pp. 2065–2069, 2018.

Index

Pages in **bold** refer to tables.